AF594002

IN THE TRACKS OF THE YETI

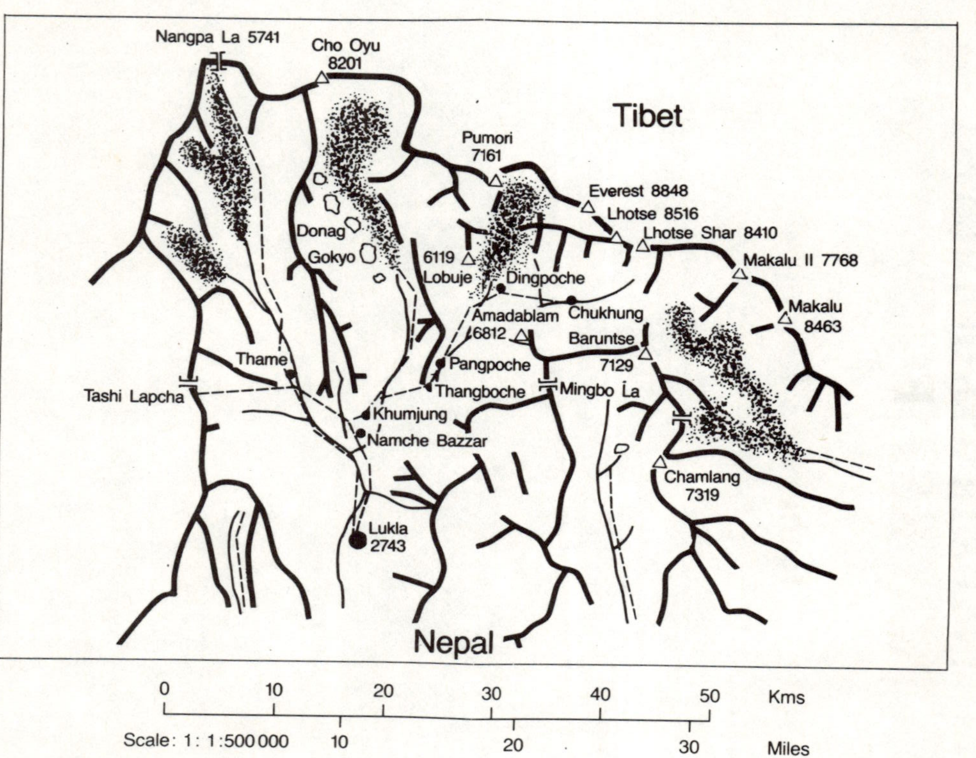
Tibet
Nangpa La 5741
Cho Oyu
8201
Pumori
7161
Everest 8848
Lhotse 8516
Lhotse Shar 8410
Makalu II 7768
Makalu
8463
Donag
Gokyo
6119
Lobuje
Dingpoche
Chukhung
Amadablam
6812
Baruntse
7129
Pangpoche
Thame
Tashi Lapcha
Thangboche
Mingbo La
Khumjung
Namche Bazzar
Chamlang
7319
Lukla
2743
Nepal
0
10
20
30
40
50
Kms
Scale: 1: 1:500 000
10
20
30
Miles

IN THE TRACKS OF THE YETI

Robert A. Hutchison

Macdonald

A Macdonald Book

First published in Great Britain in 1989 by
Macdonald & Co (Publishers) Ltd
London & Sydney

British Library Cataloguing in Publication Data

Hutchison, Robert A., 1938–
In search of the Yeti.
1. Nepal. Description & travel
I. Title
915.49′604

ISBN 0-356-17942-7

Typeset in Times Roman by Leaper & Gard Ltd
Printed and bound in Great Britain by
Biddles Ltd, Guildford and King's Lynn

Macdonald & Co (Publishers) Ltd
66–73 Shoe Lane
London EC4P 4AB

A member of Maxwell Pergamon Publishing Corporation plc

ACKNOWLEDGMENTS

Many people helped make the Yeti Research Project possible; they cannot all be named but among those who deserve special thanks are Nissam Marshall in Paris, Babeth Mary, Alex Souto and Bob Walz in Geneva, Michael Brown and Arthur Radley in London, Dave Wilkinson in Birmingham; and Bernie O'Connell in Leysin. Daughters Tamara and Chloe brought order to my secretariat. David Wileman, chief librarian of the Royal Geographical Society, was untiring in helping locate research material. The doctors Thierry Wälli and Pascal Gertsch of Les Diablerets gave willingly of their time and advice in organizing the medical side. Stamos Sports in Argentière, France, knew the ins and outs of the latest equipment and were generous suppliers. Ken Rawlinson of Phoenix Mountaineering enthusiastically provided six tents that gave welcome shelter during cold nights. But the project would never have gone forward if Dominique Clavien, a most understanding banker at Monthey in the Swiss canton of Valais, had not shown faith when the darkest moments were at hand.

DEDICATION

This book is dedicated to the memory of Eric Shipton, a source of inspiration to anyone who has read his works and knows the territory. He began the Snowman myth in its modern form with his photographs taken on Menlungtse Glacier in October 1951. His integrity was put in doubt and his character maligned because of them. But his sincerity speaks so eloquently in his description of the encounter that it is worth quoting:

> It was on one of the glaciers of the Menlung basin, at a height of about 19,000 feet, that, late one afternoon, we came across those curious footprints in the snow, the report of which has caused a certain amount of public interest ... We did not follow them further than was convenient, a mile or so, for we were carrying heavy loads at the time ... I have in the past found many sets of these curious footprints and have tried to follow them, but have always lost them on the moraine or rocks at the side of the glacier. These particular ones seemed to be very fresh, probably not more than 24 hours old. When Murray and Bourdillon followed us a few days later the tracks had been almost obliterated by melting. Sen Tensing, who had no doubt whatever that the creatures (for there had been at least two) that had made the tracks were 'Yetis' or wild men, told me that two years before, he and a number of other Sherpas had seen one at a distance of about 25 yards

at Tengboche. He described it as half man and half beast, standing about five feet six inches, with a tall pointed head, its body covered with reddish brown hair, but with a hairless face ... Whatever it was that he had seen, he was convinced that it was neither a bear nor a monkey, with both of which animals he was, of course, very familiar. Of the various theories that have been advanced to account for these tracks, the only one which is in any way plausible is that they were made by a langur monkey, and even this is very far from convincing, as I believe those who have suggested it would be the first to admit.

(*Everest* 1951, The Mount Everest Reconnaissance Expedition, Hodder and Stoughton Limited, 1951.)

A Note about Spellings:

As Sherpa-ka is a spoken rather than written language, there are many spelling variations for proper nouns and place names. For the most part, though in a few cases not, I have relied upon the place-name spellings used in the Schneider maps of Tamba Kosi, Shorong/Hinku and Khumbu Himal.

CAST OF NEPALI AND SHERPA CHARACTERS

Gyalzen Sherpa	My friend and host from the village of Pangpoche.
Kanche	Gyalzen's wife.
Pasang Tshering	Their eldest son.
Pemba Kunga	Second son.
Mingma Yanchin	Eldest daughter, also known as Yangtse.
Tseden	Fourth child, whose behaviour earned her the sobriquet of 'the Tyrant'.
Mingma Ramu	Youngest daughter.
Ang Dorje (Pangpoche)	Gyalzen's next-door neighbour and boyhood friend.
Ang Dorje (Khumjung)	deaf-dumb Sherpa who came across a yeti drinking from a stream near Dolle, hit it with his stick and died 18 months later.
Ang Dahki	Sherpani from Pangpoche who said she heard a yeti scream near Chhukhung.
Ang Kanchi	Sherpani paramedic and Tengboche electrician.
Ang Rita	climbing sirdar from Thame who reportedly lost a yak to a hungry yeti near Phurte.
Ang Zumbu	Macherma lodge owner, a stickler for cleanliness.
Baniya, Beg Bahadur	National Park auxiliary working with Bijaya Kattel on the Phortse musk deer project.
Baniya, Chet Bahadur	Beg Bahadur's uncle, also working on the musk deer project.
Bir Bahadur	cook on our caravan trek from Jiri to Pangpoche.
Buddha Chendin	16th-century holy man, Sanga Dorje's father.
Dakhu	Pangpoche herdsman who saw a yeti in the pastures behind Mingbo.
Dawa Namge	Gyalzen's father.
Dorje, Lama	foreman at the Phunki Drangka hydro site.

Frua Sange	Lhapka Degi's younger sister.
Guru Rimpoche	the great reformer of Tibetan Buddhism, also known as Padmasambhava.
Gurung, C.B.	Nepali Army colonel who led army reconnaissance team to Everest Base Camp.
Kaaputi	Kanche's widowed sister who lives in Bhandar.
Karma (Namche Bazar)	Kanche's father.
Karma (Lobuche)	Yanchin's husband. Together they run Above The Cloud Lodge, Lobuche.
Karma Tangi	younger sister of Ang Zumbu, the Macherma lodge owner.
Kattel, Bijaya	animal biologist and Nepal's senior wildlife officer.
Lhapka Degi	our Dolle hostess and sister of Lhapka Dorje.
Lhapka Dorje	the Khumjung skier.
Lhapka Tenzing	Gyalzen's neighbour who hosted the Lohsar pujah.
Lhaupa Dolma	Sherpani from Khumjung who was attacked by a yeti at Macherma.
Lobsang Tshering	Gyalzen and Kanche's brother-in-law from Bodnath.
Mingma	Pangpoche yak herder who encountered a yeti at Mingbo.
Mingma Tenzing	Nima Tenzing's brother, a resident of Phortse.
Mishra, Hemanta	director general of the King Mahendra Trust, Kathmandu.
Monticule Man	the presumed Donag Tsho yeti.
Nang Dorje	the richest resident of Pangpoche.
Nang Lobsang	owner of the Phunki Tenga lodge and tea shop.
Nang Wasser	Thawa overseer of Tengboche's Gompa Lodge.
Nawa	a Solu Sherpa who was our kitchen boy on the caravan trek from Jiri to Pangpoche.
Nawang Tenzing Jangpo	High Lama (abbot) of Tengboche Monastery.
Nima	Kanche's brother, who runs a Kathmandu trekking agency.
Nima Tenzing	Bijaya Kattel's landlord in Phortse.
Nima Wanchu	warden of Sagarmatha National Park.
Nima Yanchin	Uncle Pasang's wife and Gyalzen's younger sister.
Nursang	Macherma herder whose yaks were attacked by a yeti.
Omai	a sirdar with Yeti Mountaineering & Trekking, Kathmandu.
Parajuli, Bharat	managing director of Yeti Mountaineering & Trekking, Kathmandu.

Pasang	climbing sirdar and owner of Pasang Lodge, Namche Bazar.
Pasang Dawa	Uncle Pasang's eldest son.
Pasang Kami	owner of P.K. Lodge, Namche Bazar.
Pasang Sherpa	also known as Uncle Pasang, a Pangpoche yak herder and Gyalzen's brother-in-law.
Pasangma	Gyalzen's mother.
Paudel, Ghanashyam	Bharat Parajuli's nephew and partner in Yeti Mountaineering & Trekking.
Paudel, Saroz	Ghanashyam's younger brother.
Pemba	Thyanboche Lodge manager.
Sanga Dorje, Lama	patron saint of the Sherpas.
Sonam Girmey	trekking sirdar and owner of Namche Bazar Lodge.
Sunderay	climbing sirdar from Pangpoche, five times atop Everest.
Tashi Gompu	grandson of Tenzing Norkey, conqueror of Everest.
Tashi Janbu Sherpa	director of Everest Trekking, Kathmandu.
Tenzing Tsherpin	owner of Thawa Lodge in Namche Bazar, keeps a mummified yeti foot in his chapel.
Tinly Lundub Sherpa	Tengboche monk and artist.
Tshomtomba	mythical tyrant king who pursued Sanga Dorje.
Tsosang	old woman of Pangpoche who wanted the yeti left alone.
Yanchin	owner of Above The Cloud Lodge, Lobuche.

CONTENTS

Introduction

The yeti is alive, though perhaps not so well, and living in Khumbu, the Sherpa homeland of the north-eastern Nepal. This strange and mysterious creature may be living elsewhere, but I can affirm that it is in Khumbu because I have seen its footprints and know where in the region it resides. A victim of the deteriorating environment, however, the yeti is on the brink of extinction and may disappear from the Himalayas before science admits that it ever existed.

But what difference does it make whether the yeti is real, the world's most endangered species? The answer should not be surprising. Every animal known to science has taught us something about our planet's natural history and evolution. If the yeti is a large primate, as its tracks suggest, then it, too, has something to tell us. It may be part of the missing link with our primitive ancestors, or it may be the bridge between prehistoric and modern apes.

This is only one aspect. From the yeti can come clues about unknown features of our own biology and anthropology. Myra Shackley, lecturer in archaeological science at the University of Leicester and, with Dmitri Yurevich Bayanov of Moscow's Darwin Museum, a leading proponent of the relict hominid theory, summed it up when she wrote that confirmation of the yeti's existence 'would have immense repercussions in studies of primate evolution alone, to say nothing of its anthropological importance to the study of myths and folklore.'[1]

The fact that the yeti is not already counted in the inventory of the animal kingdom is astonishing. It can only be a matter of time before someone photographs it and the world will be stunned, not because it is there, and living, but because in this age of space travel and spy-in-the-sky satellites it took so long to document its existence. Science, after all, has ridiculed the idea that an unknown primate could be hiding in the shadows of the world's highest mountains.

By deciding what is real and what is not, science has become the arbiter of reason in our society, assuming a role held in past times by priests and philosophers. It seems that whole reputations in the scientific community have been staked on denying the yeti's existence, or on claiming that it is something else entirely, like a transient bear or wayward macaque. When asked what is the basis for these affirmations, the answer invariably comes back that large mammals cannot survive at such high altitudes. I can only assume that the people who made such 'scientific' judgments are unaware that yaks, the long-haired Sherpa oxen with curved outward-spreading horns, graze in the summertime at altitudes of up to 5,500 metres (18,100ft).

The yeti, any Tibetan or Sherpa will tell you, lives in virtually the same climes as the yak — between the rhododendron, birch, spruce and bamboo forests around 3,400 metres (11,170 ft) and the gentian-speckled pastures above 5,000 metres (16,425 ft). In fact, I will go several steps further: there is not just one type of yeti-like animal roaming the high Himalayas, but certainly two, and perhaps three unknown and unclassified species that the Sherpas and other Himalayan people generally lump together as the yeti.

If the yeti exists, particularly if several species of yeti exist, then why, the doubters ask, have they not been sighted? Why

---

[1]Shackley, Myra: *Wildmen: Yeti, Sasquatch and the Neanderthal Enigma*, Thames and Hudson, London 1983, p. 175.

has a carcass or skeleton not been uncovered, or samples found of droppings and hair?

For the sake of this introductory argument, and so as not to complicate the issue, I also will place all species of yeti in the same basket. Of course the animal has been sighted, but not frequently, and for this good reason: it is timid, by preference nocturnal, and its population not large. In all of the Khumbu Region, I estimate its numbers to be not more than a score but perhaps as few as four or six.

Most sightings, obviously, have been by Sherpas and other mountain people. For some reason, science considers them unreliable. The first sighting by a European was recorded in 1903 by British traveller William Hugh Knight, a member of the Royal Societies Club, while returning to India from Tibet. Since then a number of Europeans have come across the animal in various parts of the Himalayas, at various times and at varying altitudes. N.A. Tombazi, a Greek zoologist employed by Ralli Brothers, Bombay, gave a good description of a yeti he saw on the Zemu Glacier in Sikkim in 1925, and British climber Don Whillans saw one scamper along a ridge in the Annapurna Sanctuary in 1970. More recently, in 1986 Italian climber Reinhold Messner saw two on the same day while walking through the forests north-east of Everest. In most cases the sightings were fleeting and from a distance.

At various times, carcasses have been found, and on at least a couple of occasions a yeti-type animal has been shot by Russian or Chinese soldiers. Myra Shackley tells us that villagers from Tharbaleh, near Rongbuk, on the Tibetan side of Everest, reported seeing a drowned yeti some years back. It had been washed onto the rocks of a mountain river near their village. They described it as about the size of a small man with a pointed head and covered in reddish-brown hair.[2]

Some time in the early 1970s the French ethnologist A.W.

---

[2] *Idem*, p. 64.

Macdonald met a merchant from Manang, in northern Nepal, who had just crossed a well-travelled pass from Tibet. The Menangi told Macdonald that at the Chinese border post he had seen the body of a large ape-like animal lying by the roadside. The Chinese border guards told him they had shot the animal as it ran through the woods that same morning.

There have also been reports that a mummified yeti existed in a Tibetan monastery. A merchant in the Sherpa trading centre of Namche Bazar has among his treasures what he claims is a mummified yeti foot taken from a Tibetan gompa sacked by the Chinese. I have seen it twice now and it is something quite extraordinary. The merchant would not let me photograph it, let alone have it X-rayed at the nearby Kunde hospital, but he was willing to sell it for $5,000, a sum not often found in my pocket.

As for hair, in 1979 a Sherpa from Khumjung came across a yeti drinking at a stream near Dolle, one day's trek to the north of Namche Bazar, and to frighten the animal away he hit it with his stick. A tuft of hair remained on the end of the stick which French 'yetiologist' René de Milleville was able to retrieve and send to the Musée d'Histoire Naturelle in Paris for analysis. According to anthropologist Jean-Jacques Barloy, the hair belongs to a large unknown primate. Supposed yeti droppings were collected by the 1954 *Daily Mail* expedition in the upper Gokyo valley, north of Dolle.

The alleged yeti scalps at the Khumjung and Pangpoche gompas are interesting, but given their ceremonial role during the bacchanal summer *dumji* festival I tend to believe they are Sherpa reproductions. As for the absence of bones and teeth shards, one must understand the climatic extremes that reign in the high Himalayas, and that predators or carrion-eaters pick even the bodies of climbers clean, while the elements reduce bones to dust within two or three years. How many climbers have disappeared without trace, starting with Mallory and Irving? No, it is not surprising that bones have never been found.

Another question frequently asked is 'Why have all previous yeti expeditions failed?' I believe the answer lies in a combination of several or all of the following points:

1. The expeditions were too large and opulent;
2. They were inappropriately equipped;
3. They had a poor understanding of the animal and its psychology;
4. They were not static enough;
5. Their members were inadequately prepared for the physical hardships.

If, as I maintain, the yeti exists, particularly if several different types of yeti exist, then what kind of animals are they? Two candidates most frequently advanced are the Himalayan or Tibetan bear and an anthropoidal ape that has either wandered into the higher-altitude pastures, forests and snowfields or, as new slopes are de-forested and opened to agriculture in the lower valleys, it has migrated to higher, more remote regions and adapted to a colder alpine environment. This is entirely possible, in which case the yeti would not be a new species but a misidentified member of an existing species, or perhaps a slightly modified genus of that species.

But two other candidates occasionally mentioned are more intriguing. The one that most vividly captures our imagination is Neanderthal Man. His fossilized remains were first discovered by limestone quarriers in a cave in the Neander Valley, east of Dusseldorf, in 1856. Scientists have determined that Neanderthal Man was a citizen of the Middle Paleolithic age, and until recent discoveries in Israel he was generally considered to be a link between *Homo erectus*, who left Africa some 1.4 million years ago, and *Homo sapiens*, that is to say modern man.

Neanderthal Man, it now seems, lived only in Europe, western Asia and, for a relatively short time, in the Middle East. He did not evolve into a higher hominid but remained a distinct species of Man who was soon overtaken in the evolutionary race by archaic *Homo sapiens*. Neanderthal Man was short in stature, averaging a tad more than five feet in height. His thigh bones were slightly curved, indicating that he walked upright, but only just. He had an exceptionally long and full face with heavy brow ridges, a flat, retreating forehead and a chinless but robust jaw. His body was powerfully shaped, with a deep chest. His arms were gangly and muscular and he had a brain cavity nearly as large as ours, except that a smaller portion of the Neanderthal's brain occurred in the frontal lobes. All this gave him a somewhat simian appearance.

In the decades following the initial Neanderthal discovery, so many other skeletal remains of this type were found in caves in Europe, the Middle East, and Asia that Neanderthals became popularly thought of as cave men. But something happened to them after the climax of the last ice age — that is to say about 38,000 years ago — for they suddenly disappeared. About this time, a new race of humans appeared, moving out of Africa along the Syro-African rift into Eurasia. They were *Homo sapiens sapiens*, an evolved form of archaic *Homo sapiens*, more highly cultured, for they lived in larger groups of more than one family and were skilled nomadic hunters who exploited an effervescing technology. They developed more sophisticated weapons, created jewelry, drew polychrome paintings, and cooked in stone-lined pits, for they knew fire. *Homo sapiens sapiens* is sometimes referred to as Cro-Magnon Man, for his remnants were first discovered by railway workers at Cro-Magnon in the Dordogne valley of south-western France in 1868. He was tall — probably close to 1.85 metres (6 feet) — and muscular, with straight thigh bones. His head was balanced as in modern man, with a high forehead, a well developed chin and an absence of the strong brow ridges that characterized the Neanderthals.

One school of palaeontology holds that when Cro-Magnon Man first appeared, being more intelligent, more agile with weapons, with a higher social order, and above all more aggressive than the thickset, rather placid Neanderthal, he pursued, persecuted and murdered his less evolved cousin into extinction. It is a fact that the Neanderthals disappeared from the face of the earth rather abruptly. The proponents of the relict hominid theory believe the statistical probability is high that some Neanderthals survived by retreating to the most inaccessible habitats imaginable, like mountainous wastes or the barren steppes of Outer Mongolia where they adapted regressively to the hardier environment and live to this day.

This is an attractive theory for a number of reasons. In the vast Mongolian expanses a type of wildman seems to exist. He is known locally as the *almas*, and logically he could be a Neanderthal survivor. Almas may have wandered south to the mountainous regions of China, where they are known as *yeren*. Almas or yeren are rarely seen, but from time to time the local populations do come across them. In all likelihood a few even crossed the Himalayas by the traditional migratory routes, which would explain one set of footprints I observed at Macherma in January 1988, as well as footprints seen in the upper Arun valley by Oregon-based safari organizer Peter Byrne, field director of the 1958 Thomas Slick yeti-hunting expedition, and possibly the figure seen by William Hugh Knight near Gangtok, months before the 1904 Younghusband raid into Tibet.

The other candidate for the yeti is a throwback to the supposedly extinct ape *Gigantopithecus*, which evolved in Asia at least nine million years ago. The morphology of *Gigantopithecus* is known only from a few fossilized teeth and jaw bones found in China and India, but apparently it was a ground-feeding ape about the size of the modern gorilla, with molar teeth well adapted to crushing tough material. It is supposed to have died out half a million years ago, but its descendants could still survive, like Neanderthal Man, in some inaccessible habitat.

Now all this is not as far-fetched as it sounds, and for very good reason. The Himalayas are geologically the result of a collision between two continental plates: the main Asian landmass and the Indian subcontinent. Because these two plates have been thrust together, the fauna of each plate intermingle along the point of contact, which is the Greater

Himalaya chain, making the area extraordinarily rich in wildlife. Its valleys are not easily accessible, even with modern transport, and species previously unknown to science are regularly discovered there. Recently the French anthropologist Michel Peissel reported that a giant panda had been found in southern Tibet, near the border with Bhutan. Within the last few decades a new genus of red panda was discovered in central Nepal. To the north-east of the Himalayas and in Bhutan primeval forests are still untouched by man and contain surviving species of tree from the Tertiary period.

'It is possible that not only flora but also some of the fauna of previous times could have survived,' Chinese anthropologist Zhou Guoxing, vice-director of the Beijing Museum of Natural History, wrote in a 1982 paper.[3]

Dr Zhou, who is active in primatology research in China, believes that two yeti species exist in regions bordering on the northern aspects of the Himalayas. He thinks the smaller one may be an unidentified species of macaque. But the larger version has 'possible affinities to the fossil ape *Gigantopithecus*'.

If confirmed, this would be a capital discovery. Large, tailless anthropoidal apes — the chimpanzee, orangutan, gibbon and gorilla — are very manlike in their anatomical structure. A century ago the English biologist Thomas Henry Huxley (1825–95), an early exponent of Darwinism, concluded that 'whatever system of organs be studied, the comparison of their modifications in the ape series leads to one and the same result — that the structural differences which separate Man from Gorilla and Chimpanzee are not so great as those which separate the Gorilla from the lower apes, that is, monkey.' This proposition became known as the 'pithecometra thesis'.[4] More recently it was reinforced by the discovery of similarities in certain blood groups of anthro-

---

[3]Zhou Guoxing: *Wildman Research in China,* Cryptozoology 1, 1982.

[4]T.H. Huxley: *On Man's Place in Nature*, London 1863.

poidal apes and man, and a shared susceptibility to specific diseases. So whatever it is, ape or wildman, the yeti can perhaps teach us something more about ourselves.

In one form or another wildmen and yeti have been around for a long time. In fact, throughout history every mountain people, or people who live in the near vicinity of mountains, have had them in their culture. Human imagination works in such a way that very little is actually, at its origin, imagined. Most legends have, or had somewhere back in the cobwebs of time, a factual base. Anthropologists excavating the El Juyo cave, in the foothills south-west of Santander, Spain, found a 35-centimetre (14 inch) stone head depicting a creature half-man and half-beast. The El Juyo cave was frequented by a Cro-Magnon tribe 14,000 years ago and the carving could have been their rendering in stone of a yeti or wildman.

My feeling is that Neanderthal or *Gigantopithecus* relicts did exist, and may still, in order to have inspired carvings like the El Juyo head or legends such as the Epic of Gilgamesh. This Babylonian odyssey contains the earliest reference in literature to a wildman. Written in cuneiform characters across twelve clay tablets something like 4,000 years ago, the epic was found in King Ashurbanipal's library at the Assyrian royal city of Nineveh. The tablets tell us that Gilgamesh was a mighty Sumerian king who befriended Enkidu, a creature described as half man and half beast. Enkidu first lived among the animals but was tamed and moved to the cities of man. Gilgamesh admired Enkidu for his superhuman strength and immense bravery. The epic describes their voyage together through many lands in search of the key to immortality.

In Greek and Roman imagination, the wildman was depicted as a satyr who became the consort of Dionysus, the god of wine. The satyr was eventually stylized as a cloven-footed creature with a man's head and torso. But in early Greek drawings he was entirely humanlike, except for a tail, with a broad face, long hair and straight limbs. Then Alexander the Great added a new wrinkle to the wildman myth when, in 327 BC, while camped beside the Indus River, scouts reported to him that a 'Himalayan tribe whose feet are turned back to front' lived in the land of the snows. The scouts claimed they were unable to bring any of these snowmen back to Alexander's camp because the creatures could not breathe at lower altitudes.[5]

Medieval Europe knew varying forms of wildmen. The masks of hairy monsters paraded at carnival time through Swiss Alpine villages are a reminder of them. In the canton of Valais people still talk, though jokingly, of a hairy nocturnal creature that supposedly roams the forests in the valley of Saas Grund, scaring the local inhabitants.

Appalling presences have haunted the Alps since modern man first arrived there. Until the sixteenth-century villagers in the four Swiss forest cantons solemnly followed their priests to nearby lakes and pitched rocks into the water to frighten away unwanted monsters. Alpine monsters were a favourite subject of Johann Jacob Scheuchzer (1672–1733), a noted professor at the University of Zurich. He collected the sworn testimony of supposedly honest men who said they had seen them, and concluded that Alpine monsters varied greatly in appearance.

Richard Bernheimer, in his famous book *Wild Men in the Middle Ages* (Harvard University Press, 1952) describes a fourteenth century psalter, found in the British Library, that belonged to Queen Mary which shows a wildman being harried by a pack of dogs. The wildman pictured in the psalter bears a remarkable similarity to Sherpa descriptions of one type of yeti. But we also have examples of wildmen carved in stone over the doorways of Gothic churches, illuminations of wildmen in medieval manuscripts, medieval paintings of wildmen and wildmen represented in medieval pageants. Anthropologist Ivan Sanderson did a survey of wildman representations in European literature and found them prevalent from the eighth to the sixteenth centuries. Breughel the Elder's rendering of the Battle of Carnival and

[5]Robin Lane Fox: *Alexander the Great*, Penguin, 1973, p. 333.

Lent shows the carnival emperor and his followers closing in on a villager dressed as a wildman in a leafy, green costume.

We owe our system of modern biological nomenclature to the Swedish naturalist Carl Linnaeus who in 1758 published his *Systema Naturae* cataloguing all known animal species. In it he named several examples of living man: *Homo sapiens* (ourselves), and a primitive man whom he called *Homo troglodytes*, because he lived in caves. These humanoid species he classified, with living apes, as primates. He based his conclusions about existing troglodytes on reports by contemporary authors and travellers who claimed either to have seen them or who had been told of their existence. But Linnaeus's theory that primitive man had survived was derided by science.

Twenty-six years after Linnaeus published his work, *The Times* of London reported that Indians at Lake of the Woods, in the Canadian province of Manitoba, had captured a 'huge, manlike, hair-covered creature'. It apparently did not adapt to captivity and died.[6] Another hundred years passed, Neanderthal man was unearthed, and then employees of the British Columbian Express Company captured a creature they called Jacko. They described Jacko as half-man, half-beast,[7] one supposes somewhat akin to ancient Enkidu or one of the species of modern yeti. Both the Lake of the Woods and British Columbian creatures were undoubtably sasquatches, the North American cousin of the yeti whose ancestors probably wandered across the 83 kilometres (52 miles) of Bering Strait when the ocean level was lower and a land bridge brought Asia and North America together.

The Caucasus and Pamirs also have a history of wildmen and yetis. Odette Tsherine in her book *The Yeti* relates the story of Zana, a female wildman who in the nineteenth century was captured in the forests of Mount Zaaden in Georgia and lived domesticated in the village of Tkhina, 125

---

[6]Shackley, *op. cit.*, p. 35.

[7]*The Colonist*, a Victoria, B.C. newspaper, July 4, 1884.

kliometres (78 miles) from the Abkhazian capital of Sukhumi, on the Mokvi River. Her skin was said to have been greyish-black and covered with reddish hair. She was capable of making inarticulate cries and learned to obey simple orders. She would eat anything offered her but was especially fond of meat and wine. She could outrun a horse and swim the Mokvi River when in flood. She had a sharp sense of hearing and cowered when her master shouted at her, but otherwise became proficient at menial domestic tasks such as grinding grain into flour. Zana, more incredibly, became the mother of half-breed children: two sons and two daughters survived and they grew into capable citizens engaged in normal work and social life. She died in the 1880s.[8]

When considered in this global context, then, the yeti encounters that Sherpas have experienced and the footprints that British climber Eric Shipton and others have seen in the high Himalayas of Tibet, Nepal, Sikkim and Bhutan are not so unexplainable after all. Determined to unravel a bit more of the mystery, I travelled to Nepal in the autumn of 1987. My intention was to find and track the yeti to its lair and if possible not only to photograph it but to study its nesting, mating and eating habits while obtaining samples of its hair, excretion and other materials so that a biological study could be begun and a strategy drafted for its survival. For five months I pursued this goal, an intruder along the frontier of the unknown. Although my winter travels in Khumbu did not produce the hoped-for photograph of a yeti, it resulted in what I believe is a wealth of new evidence about the animal's existence and living habits.

It was only on a visit to Israel in March 1989, well after my return from Nepal, that another piece of the yeti puzzle fell into place. To explain its importance, it is first necessary to digress for a moment, as some of the most recent archaeo-

---

[8]Odette Tsherine: *The Yeti*, Neville Spearman, London 1976, pp. 155-157.

logical evidence uncovered in the Middle East is in contradiction with existing dogma about Man's evolution.

After visiting archaeological sites in the Jordan Valley, the world's oldest migratory axis, I met at Hebrew University in Jerusalem a palaeontologist who is a leading proponent of a theory known as 'Noah's Ark'. He is Prof. Eitan Tchernov, a person of boundless energy and enthusiasm, an extraordinary detective following Man's dusty evolutionary route through the eons of prehistory. Tchernov maintains that the latest genetic studies and methods of dating prehistoric material support the proposition that Man evolved in one place, Africa, and that we all share a common ancestor: one group of *Homo erectus*, who replaced *Homo habilis* about 1.5 million years ago. Members of this *Homo* genus started to occupy more versatile habitats and experimented with a wider spectral array of food resources. This was in contrast with *Homo erectus*'s remote ancestor *Australopethicus* and other early primates.

Man underwent further socialization during the development of *Homo erectus* and since then has always lived in groups, at first in the Great Rift Valley of eastern Africa. Moreover, *Homo erectus* was a rapid breeder and an adventuresome traveller. Soon his offspring spread across Africa and then followed the Great Rift north-eastward along the narrow Middle Eastern passageway that Tchernov calls the Levantine Corridor. Within a relatively short period, they had populated the Eurasian and African continents, from Abidjan to Vladivostok.

Man's first stop out of Africa, as far as we know from the latest archaeological evidence, was a lakeside plain that in recent times was called Ubeidiya. The lake dried up long ago and today Ubeidiya is little more than a tomato patch, albeit a rather large one, located in the Jordan Valley, a few kilometres south of the Sea of Galilee. The Jordan Valley, as part of the Syro-African rift system, was then as now the main crossroads between Africa and Eurasia. Prof. Tchernov, who spent fifteen years excavating the uplifted layers of strata caused by tectonic faulting at the Ubeidiya site, says that *Homo erectus* stopped there about 1.4 million years ago. At this stage, Man most probably knew neither fire nor clothing; he fashioned only the most primitive stone tools, was a poor hunter, relying on bigger predators to slaughter food for him, did not bother to construct dwellings and scorned even the crudest shelters except, perhaps, in times of exceedingly bad weather. He lived in small groups among animals, as one with them.

One day at Ubeidiya, Man had a picnic by the lakeshore. Prof. Tchernov has studied the fossilized remains of this picnic. It seems that a sabre-fanged tiger, long since extinct, killed a baby hippopotamus and dragged it onto the beach for a meal. After stripping away its very best parts, and gnawing on a few bones, the sabre-fanged tiger departed for an afternoon nap. Along came Ubeidiyan Man with his family and finished the meal. Using stone scrapers and choppers they dissected what remained of the baby hippo and ate its flesh. They used rocks to break open its bones and then sucked out the marrow. The sabre-fanged tiger before them had also done this, but much less artfully, leaving its fang marks on some of the bones.

As Ubeidiyan Man had not yet developed his own arsenal of weaponry that would permit him to kill big game, he had to be content with the sabre-fanged tiger's leftovers, the second best parts. The bones that Man scavenged from the remains of the carcass bore human teeth marks. Amongst the bones, he left some of his tools, a few having been broken during the undressing of the hippo. They are the most ancient implements ever found outside Africa. Other than teeth marks, the bones were scarred by the very same scrapers Ubeidiyan Man discarded at the site.

In his wanderings over the next few hundred thousand years, *Homo erectus* changed slowly. Now we are at the heart of the Noah's Ark theory. It holds that in Africa a new genus — archaic *Homo sapiens* — emerged from *Homo erectus* about 200,000 years ago, before the nascence of Neanderthal Man, whose remains, according to Tchernov, have never been found in Africa. Another 100,000 years later, archaic

Homo sapiens moved into the Levantine Corridor, pushing northward until he reached the southern limits of the Ice Age glaciers in Lebanon and Syria. In Europe, meanwhile, *Homo erectus* had evolved gradually into Neanderthal Man, thick-set and better adapted to the colder climate.

Of the two sub-species, Neanderthal Man had the more difficult lot. As the glaciers advanced in Europe he found it increasingly difficult to eke out a living. He moved southward, entering the Levantine Corridor from the north about the same time as archaic *Homo sapiens* was edging out of Africa along the Sinai and Jordan land bridges. One Franco-Israeli group of palaeontologists, of which Prof. Tchernov was a member, found the remains of such a Neanderthal in a cave along the Mediterranean coast. Other fossilized Neanderthal remains were found elsewhere in Israel. Curiously, it seems that for a while archaic *Homo sapiens* lived side by side with his Neanderthal cousin. Being larger and stronger than archaic *Homo sapiens,* Neanderthal man was able to exclude this more puny creature from the best feeding sources.

Unable to compete, *Homo sapiens* retreated back to Africa during the first part of the Glacial period. Something happened to him there. He developed over the next several thousand years new skills and technologies which he applied to hunting and food gathering, in the process creating for himself an entirely new culture. About 40,000 years ago, now known as *Homo sapiens sapiens,* he again stirred out of Africa along the same routes, coming into contact once again with the stronger but less nimble Neanderthaloids. In his new guise, *Homo sapiens sapiens* became known as Cro-Magnon man, this name, as already mentioned, having been derived from the site in south-western France where his remains were first found. Very quickly he removed Neanderthal Man from the landscape by the simple process of *competitive exclusion,* not by 'tooth and claw'.

Because he exploited nature's resources better, and took over the best feeding points, Cro-Magnon Man signed the warrant for Neanderthal's extinction. In a few thousand years Cro-Magnon had exterminated Neanderthal and with incredible speed repopulated the Middle East, Asia and Europe.

Prof. Tchernov had found in me an avid listener. I was fascinated, particularly by Ubeidiyan Man's lakeside snack. His feeding and living habits were exactly consistent with the Sherpas' descriptions of the habits of a yeti. When I mentioned this to Tchernov, he became interested. Like *Homo erectus*, the yeti is also a scavenger, living to an extent off the kills of other animals, notably the snow leopard. This close hunting/scavenging relationship between the snow leopard and the yeti was uncovered by Prof. Boris Porshnev, of the Soviet Academy of Sciences, during his research into the relict hominid theory. In the 1950s Prof. Porshnev headed a Soviet 'Snowman' Commission and during his research he learned from Kirghiz nomads in the Pamirs about the yeti's shadowing of the snow leopard. Thirty years later I heard similar stories from the Sherpas of Nepal's Khumbu region.

Like *Homo erectus*, the yeti does not know fire, speaks only with guttural sounds, and lives in no permanent dwelling. To my mind, these parallels add fuel to the hypothesis, first scientifically explored by Prof. Porshnev in his 1960 work, *The Contemporary Situation as Regards the Problem of Relict Hominids* (only 180 copies published) that the yeti is really the last survivor of our prehistoric anscestors, pushed into adapting to a remote and uninviting habitat, his ultimate refuge from Cro-Magnon man, that is to say ourselves.

'You have described for me the perfect behaviour of a hominid scavenger,' Prof. Tchernov said after I finished explaining the Sherpas' portrayal of the yeti's lifestyle. 'I suppose it is possible, though unlikely, that prehistoric man could have survived in a remote environment if sufficient food existed there for him to eat.'

Wanting to convince Tchernov that there was adequate food in the high Himalayas for a mammalian scavenger, I mentioned the snow leopard stalked and killed partridges and snow cocks. 'Too small,' he said. 'Not substantial enough.'

'How about pikkas and voles?' I tried. He shook his head. 'And Himalayan tahr?'

'You mean the big mountain goat? Then it's possible,' he replied.

'There are also musk deer, and, of course, yaks,' I added.

'Does the snow leopard kill yaks?'

'Probably. But the Sherpas always blame it on the yeti.'

'Then there's no doubt that he would have enough food to survive.'

Prof. Tchernov confirmed that the discovery of a yeti, even if it proved not to be a relict hominid but a lesser primate that had devolved to tundra living in the most hostile environment in the world, would be of capital importance. 'It could explain so many things,' he said. 'But science is full of unproven hypotheses. What we need is proof. Bring me proof.' He agreed that the mummified foot I had uncovered in Namche Bazar would be a beginning.

While interest in the yeti was the primary reason for my travels along Nepal's frontier with Tibet, there was one other, prompted by a journey I had made through the Garhwal Himalaya of north-western India with the Indian ecologist, Sunderlal Baguhuna. A disciple of Gandhi, Baguhuna gave up power broking in Indian politics to fight for the rights of the Garhwali hill people and for the protection of the Himalayan forests from rapacious timber merchants, promoters of monster dam projects and unscrupulous mine operators. Baguhuna's efforts led to the founding of the Chipko Movement which has since spread worldwide and seeks to save all forests from becoming fodder for the mindless expansion of industrial society. Baguhuna revitalized Gandhi's theories on the economics of development and translated them into today's context.

In the spring of 1987 we went on a ten-day walk together in Tehri-Garhwal and I learned more in that short time about the problems of the Third World and our deteriorating environment than I had during the previous ten years. Baguhuna's thesis, which he owes to Gandhi, is that if we wish to avert the worst crisis mankind has ever faced we must rein in the forces of uncontrolled materialism. The quality of life in the industrialized nations cannot continue to progress at the expense of the Third World, and the cult of mass consumption must end if a devastating North-South conflict is to be avoided.

Baguhuna says that the wealthy nations are raping the resources of the Third World to maintain a wasteful, hedonistic society, and that this is condemning 90 percent of the world to live in constantly degrading conditions in order to maintain the high standards of the other 10 percent. One of the manifestations of this problem, albeit a rather minor one, is the trekking industry in Nepal. While it brings immediate benefits to the mountain people, they are mortgaging the well-being of future generations by plundering their forest resources to provide cooked food, warm lodging and hot showers for trekkers from the more leisured societies of this world. One of the victims of this rape happens to be the yeti, but the people themselves will suffer unless a few basic solutions are found.

Essentially, Baguhuna's global message is that when the societies of waste, through their wealth of pollution, start to imperil the air we breathe and the very atmosphere in which we co-exist, and when their affluence starts modifying our climate, bringing drought followed by freak storms, desertification and changing monsoon patterns, then the Third World will have but two alternatives: fight or die. Better then to reduce the excesses of mass consumption and lower the leisured societies' expectations of greater wealth and ease of living to avert this cataclysm and try, in a measured way, to close the gap between the developed world and the Third World in both living standards and quality of life.

This book offers no solutions. It attempts to describe conditions encountered during a winter stay in the high Himalayas in hopes that a growing number of people will become aware of the problems and seek to eliminate them. Otherwise there is every chance that, as a first step to our own destruction we are destined to complete the work begun 38,000 years ago by Cro-Magnon Man and finally bring about the extinction of the yeti.

Chapter 1

Kathmandu

Kathmandu, city of the gods, is the capital of Nepal, the world's only Hindu kingdom, a place where the cultures of men and the frontiers of time meet and frequently clash. Until the overthrow of the Rana princes in 1951, Kathmandu was a closed city, mysterious and forbidden, an oasis on the age-old trade route between the Tibetan plateau and the plains of northern India.

For most people, the journey to Kathmandu is a search for adventure, a quest for oneself, a spiritual and mystical detour. This was certainly my case when I arrived in Kathmandu's emerald valley on Saturday, 10 October 1987. But the tensions that had marked my last weeks in Europe, as I prepared for a five-month search for the yeti, had exhausted me and I was looking forward to a few days of rest before setting out for the remote Himalayan highlands, along Nepal's border with Tibet. From previous visits I should have known that rest was unlikely, as a few unsettled days lay ahead while I acclimatized to Nepal's culture warp.

Most first-time visitors are blithely unaware as they deplane from the jets that fly them into the steeply-terraced valley that the dimensions of time are quite different here. Time moves at a mixed pace, generally slower than elsewhere. To begin with, Nepalese time is fifteen minutes ahead of Indian time, which is five and a half hours ahead of Greenwich. The pressures of everyday life are also quite different, often reflecting currents not readily perceptible to westerners. If perhaps more subtle, they are no less vital, more fully embracing even, and they can be potentially explosive.

Immediately upon stepping out of the passenger terminal at Tribhuvan International Airport, travellers are confronted by a world of contrasts: of new order seeking to impose itself upon the old, of the many rivalries that mark Asia, and of the aching gulf between rich and poor. No time exists for gentle immersion. It is a head-on collision with a culture whose delicate mosaic is being eroded by Occidental aid and mass-consumer materialism.

In front of the old terminal building, the tour operators and hotel guides are waiting for their clients, sheltered from the sun by a corrugated roof, sipping milk tea (*dudh-chiya* in Nepali) out of dirty glasses. They appear unhurried, relaxed, with smiling faces that reflect the ethnic mix of the Himalayan crucible. Meanwhile, urchins tug at your bags in the hope of receiving baksheesh and three or four of them rush to hold open the doors of your taxi.

The taxis are mostly Japanese imports, worn out and dented from hard driving, lack of maintenance and the valley's rough roads. I was met at the airport by Bharat Parajuli and his nephew Ghanashyam Paudel. They are partners in a Kathmandu trekking agency appropriately named Yeti Mountaineering and Trekking. They had a clapped-out, fawn-coloured Toyota taxi waiting for me. As we wheeled away from the controlled chaos resulting from the initial encounter between the new arrivals, airport urchins, porters and tour guides, our driver complained about how slowly work was progressing on the new terminal building. It is an imposingly modern structure at the end of the airport road, faced with traditional bricks by local labourers using mostly medieval methods, under the aegis of Japanese engineers backed by Japanese financing and a few Japanese building machines.

We rattled around the traffic circle and headed for Battisputali, an eastern suburb of Kathmandu. Towards the south end of the airport's single runway, the roadway splits, right into the city, less than 3 kilometres away (1.9 miles), or left onto the Arniko Highway. Built by the Chinese, the Arniko goes to the Tibetan border, 115 kilometres (72 miles) to the north. Its narrow two lanes are only slightly less hazardous to travel than the valley's other major thoroughfare, the Tribhuvan Rajpath, which begins on the far side of the city and winds west and southward for 200 kilometres (125 miles) to the Indian border. The Tribhuvan Rajpath was built by Indian Army engineers as a gift to the Nepali people. Both roads encapsulate with a bulldozer's bluntness the geopolitics of Nepal, sandwiched between the two most populous powers on earth, each vying to dominate with aid and influence the tiny tampon-state that straddles the roof of the world. Three and a half times the size of Switzerland, Nepal has a population of 18 million, also three times larger than Switzerland's. But by Indian and Chinese standards it is uncrowded.

We veered right before the Arniko intersection onto another road that would take us into Battisputali, passing Kathmandu's only golf course, with 'greens' of rolled sand, designed by a leading American golf course architect. Further along, as we crossed the sacred Bagmati River, I could see the roofs of the Pashupatinath temple. One of the holiest Hindu sites, it attracts pilgrims from every corner of the subcontinent.

Beyond it at some distance, unseen from our road, is the Buddhist shrine at Bodnath, proudly proclaiming to be the largest *stupa* (bell-shaped repository of holy relics) in the world. Had we driven into the city, on Durbar Marg, the main thoroughfare descending from the Royal Palace, we would have seen the central mosque, a reminder that Kathmandu is a city of tolerance, where three of the world's great religions meet and worship in apparent perfect harmony.

But there is strife. From the taxi's back seat, a traveller can see scenes of an enchanted city suffering under the burden of tourist-borne inflation and modern industrial stretch. Mercedes buses on their way to and from the half-dozen five-star hotels swerve to avoid tricycled rickshaws and three-wheeler tempos (scooter cabs). Tawny-skinned street vendors in ragged clothes carry trees of flutes through the Basantapur market-place while turbanned Sikh merchants, immigrants from India, bustle past Newari-owned trading emporiums, their show windows filled with the latest Japanese transistors, cassette recorders and video cameras.

The smart Newari business class has through the ages been expert in triangular trade. Before it was largely counter-trade, that is trade by barter, but recently some Newari merchants have made a speciality out of selling bulk commodities to Tibet, using the proceeds to buy Buddhist relics and artwork that fetch a pretty price in Tokyo, and bringing back for sale in Kathmandu a broad selection of Japanese consumer products. In the backstreets of Thamel, Kathmandu's 'Latin Quarter', Tibetan refugees hawk carpets and prayer wheels to the mutitude of tourists.

Modern Kathmandu relatively sparkles, while odours of decaying garbage and open sewers cling to the walls of Thamel and permeate the narrow streets of Old Kathmandu. Gemstone merchants, tanka dealers, beggars and sidewalk sellers of phoney antiques — whether Newar, Sikh or Tibetan — are busily chasing the tourist dollar or yen, but deutsch-marks, francs or sterling will do, for everyone prefers hard currency to the King's rupees.

Bharat Parajuli and Ghanashyam Paudel had become my good friends over the previous two years of preparation for the yeti research project. Bharat was interesting because he was only two generations removed from the agrarian hill culture of middle Nepal. His grandfather came from a hilltop village to the east of the capital but had moved with his family to the city when the village well dried up and the sources of firewood and leaf fodder for livestock became ever more scarce.

The Parajulis, like a majority of Nepalis, are Hindus. Bharat received the best education Kathmandu could offer in those days, being now, in his late forties, a doctor of economics from Tribhuvan University. He has fine Indo-Aryan features, reflecting a heritage from the plains to the south, is gentle-mannered, reflective and fiercely patriotic. Bharat became a teacher and later the headmaster of a private school until one day Colonel Jimmy Roberts, who was then military attaché at the British embassy in Kathmandu, asked him to guide a party of English ladies on a trek to the central Nepali city of Pokhara. There was no road to Pokhara in 1961 and the logistics of the trip proved formidable but Bharat managed to the ladies' satisfaction. Little did he realize, though, that he was pioneering Nepal's newest industry, mountain trekking.

Jimmy Roberts retired from the diplomatic service and opened with Bharat Nepal's first trekking and mountaineering agency. They remained partners for several years, organizing the Nepal side of many early Himalayan climbing expeditions, before finally going their separate ways in the mid-1960s. Bharat married; he and his wife have two teenaged boys, both in their turn receiving the best education Kathmandu can offer today. (As a striking aside, I found in my four trips to Nepal that the growing middle class is incredibly education-oriented. A number of excellent schools exist in and around the capital and I noted that often parents would make great sacrifices to insure that their children received a better education than they themselves had been able to obtain. There is a real desire to 'become educated', as if their newly acquired knowledge and skills could lift the country out of the morass of Third World poverty. As far as I could tell this urge was paying dividends.)

Bharat's nephew, Ghanashyam, is ten years younger and a Brahmin, the highest Hindu caste. His features are broader, his complexion darker, reflecting a Gurkali influence. Ghanashyam's wife is an operations department manager for Royal Nepal Airlines; they have two children just entering school. His wife's uncle is a noted Communist, representing an electoral district east of the capital, loved by his village constituents but hated by the local landowners, and he has been arrested, allegedly for his own safety. Ghanashyam is a member of the national sports council. Bharat is joint secretary of the Nepalese Mountaineering Association.

Our taxi took us to Ghanashyam's guest house. Once

rumoured to have been a brothel, the Sukeyasu Guest House, so-named for one of Ghanashyam's Japanese friends, is now a sedate place. It hides at the end of an alleyway, behind a green metal gate, a brick house of two floors of bedrooms above a reception office, dining room and kitchen. It has eight bedrooms, a staff of five, and I find it a welcome refuge from the noise of the city. Saroz, Ghanashyam's younger brother, was waiting to greet me and we had a cup of tea downstairs at reception before I went to my second floor room for a few hours' rest.

I was by then running ten days late on my original schedule. Until my departure from Geneva roughly forty-eight hours before, it had been a cliffhanger, that is to say a touch-and-go situation, but more touch than go, as to whether I would get myself and the yeti project airborne. What had been intended as a six-member expedition was now down to two — myself and Gyalzen Sherpa, a one-time high-altitude porter, cook and yak herder from the tiny Khumba village of Pangpoche, on the pathway to Everest Base Camp. Gerard Metral, a friend from Chamonix in France, would join us in a month for a few weeks but otherwise that was, for lack of funding, the full extent of it. A Toronto marketing agency had assured me almost until the very end that a major multinational was ready to underwrite the cost of the expedition, but this, alas, fell through and there was nobody except my Swiss banker, who insisted on taking a mortgage on my 100-year-old Valaisan chalet, to provide me with financing.

By then I had discovered that mention of the yeti elicits different reactions from different people. Almost everyone with whom I had talked over the previous three years had tended to laugh away the animal as a Sherpa myth, an impossible piece of folklore. A few said 'So what?' or suggested that the 'abominable snowman' should be left alone. I had noticed that those who expressed these reactions were usually from the more materialistic segments of society. But for a minority, the idea of seeking the yeti was at once romantic and exciting, one of the last great mysteries. It was interesting to note that they were generally people from the liberal professions or associated with creative endeavour in one form or another.

For me, however, the journey into Sherpa country was quite unabashedly an escape, a search for myself as much as for the animal which I proposed to track and study. I was — and am — firmly convinced that it roams, rather desperately, through the high Himalayan pastures and valleys looking for food and shelter, a companion with whom to share its joys and hardships, and the right to live at peace with nature in the environment of its choosing, just like the rest of us. And even though the yeti was no market speculator nor a material sort, from my earlier research I had come to the conclusion that it had fallen upon hard times. This belief was in no way tied to my boyhood dream of one day testing myself in the Himalayas, of viewing the sunrise on Everest, or of finding the much-maligned snowman. Ever since seeing the pictures of its footprints taken by Eric Shipton on Menlungtse Glacier in 1951, the yeti had fascinated me.

My Toronto marketeers had been optimistic. The budget was not large — $180,000 — and they thought the advertising returns, whether I found the yeti or not, would be impressive. One of the first companies they approached was Seiko Epson Corporation of Japan, which had been using Shipton's famous close-up picture of a yeti footprint in an advertising campaign for the Epson HX-20 personal computer. Under a provocative heading — 'The spectrum of possibilities is as wide as our imagination' — some creative copywriter had strung together a hundred words that linked the poor yeti to the latest technological developments at Seiko Epson:

> Some think the idea of an Abominable Snowman is as widely impossible as his rather ungainly footprints. Others think it no more incredible than the advanced technologies, information systems and products that are already a part of our daily lives . . . So as we [at Epson] continue to create superior products that make your daily life easier, we may yet cross paths with the Abominable Snowman.

Out here, on the edge of the imagination.

This seemed the perfect invitation. So we let our imaginations roam and sent Seiko Epson a copy of the Yeti '88 prospectus. We were turned down by a rather blunt note faxed from Tokyo:

> Unfortunately we may not afford to participate in the YETI '88. We are interesting in finding out the seed of technological development. Even the seed is tiny and looks to be imaginative and unbelievable. However, it doesn't come to the participation of the Expedition of the Abominable snowman, directly.

Although the wording was jumbled, the message was sufficiently clear.

Next my marketeers tapped on the door of PepsiCo. The 1988 Winter Olympics were coming up in Calgary, and Pepsi, one of the sponsors of the Canadian ski team, became interested in adopting a light-hearted yeti rendering, produced by a friend of mine for the Yeti '88 expedition, as the ski team's mascot. In return for the rights to 'my' yeti, Pepsi said it was prepared to pay a modest sum. There was not only promise, but a smell of money in the air.

But the marketing people at Pepsi discovered that some years before the Canadian ski federation had used the yeti in an advertising campaign. The gist of the campaign was 'Don't ski like a yeti, take ski lessons.' Alas, the Pepsi sales promoters determined that this had given the yeti a bad image, like maybe it was a slouch, not chic in any event, so they declined.

Now my marketeers approached the Canadian promotional staff of Kellogg's. Suddenly there was talk of exchanging boxtops for yeti T-shirts or stuffed yeti dolls, and printing a Search for the Yeti game on the back of breakfast cereal boxes. The marketing agency, which looked forward to raking off 20 percent of the expedition's funding for their services, said Kellogg's was in the bag. At the last moment, however, Kellogg's wanted to run the idea in front of a 'children's focus group'. This, I learned, is a panel of about twenty kids who, with their mothers, are invited to an after-school party at which a dozen themes for possible promotional campaigns are presented to them. Between gulps of cola and cake they are supposed to give their opinions on each theme. This astounded me. These kids are being asked by high-powered executives to decide how a major corporation should invest between $20-30 million on a promotional campaign. I don't know what the eleven other themes were, or how and in what order they were presented, but my marketeers came back with the news that the yeti had come in at the bottom in every category of question. In contrast to Tintin, the adventuresome kid created by the Belgian cartoonist Hergé — Tintin when he went to Tibet met the yeti and was enchanted — these cola-guzzling youngsters weren't interested.

Not to worry, the marketeers said. (In fact I was worried stiff as we were then in the month of June and had been dilly-dallying around since the beginning of the year.) They mentioned that Kellogg's corporate thinkers had been known to overrule findings of children's focus groups before. And so the idea was taken to the world headquarters at Battle Creek, Michigan. The parent corporation had more money, the marketeers assured me, and surely would be interested. But the verdict, as relayed a few weeks later through the marketeers, was, 'This yeti idea is too damn cultural. We're not in the business of educating kids, but selling breakfast cereals.'

At this juncture the marketeers and I said goodbye and in desperation I contacted Richard Branson's Virgin empire. The Virgin people accepted a few T-shirts I had designed and tinkered with the idea for a while before deciding, apparently in the words of the Big Man himself, that Virgin didn't see how it could make back a $180,000 investment by associating with the yeti. So I was left with the only alternative: cutting everything to the bone and raising $48,000 in a mortgage. It was one of those nightmares which ended with a race to nowhere because you had the impression you were standing still and the whole world was slipping through your fingers. After securing the mortgage, I flew to Stuttgart to pick up an electronic night-vision scope from Carl Zeiss AG, and arranged for six Phoenix expedition tents to be carried to Kathmandu by a Channel 4 television crew from London. They were going to Nepal to film an ecological-impact production on trekking to Everest Base Camp, which had the working title of Adventures Along The Kleenex Trail. I also had to make arrangements to pick up an Everest sleeping bag from Mountain Equipment Limited at Stalybridge in the north of England, three pairs of skis from the Dynastar factory at Sallanches, near Chamonix, and to pack eleven expedition drums with clothing, gear, medical supplies and high-altitude food that would help sustain me during a winter siege in the Himalayan wilderness.

I had shipped the nine 60-litre drums, two 120-litre ones and a ski bag as air freight the day I left Geneva and theoretically they should have arrived in Kathmandu, via Delhi, one day after me. I had no extra cash for insurance, so cameras, film, gear and clothing were all at risk in the unsealed blue plastic containers. This was a supplementary worry. But at least I myself had arrived safely with $6,000 in my pocket, which was a modest beginning.

Bharat Parajuli told me that Gyalzen and his eldest son, Pasang Tshering, had come from their village of Pangpoche to meet me and had been waiting in Kathmandu since the end of September, hoping each day I would arrive, as they, too, were short of cash. Bharat had advanced them some money against my account. I arranged to meet Gyalzen at the trekking agency on the following morning so that we could begin preparations for our departure to Jiri, the road-head to the east of Kathmandu, and the walk into Solu-Khumbu, a not too accessible region in the north-east of the country.[1]

---

[1]The Sherpa homeland is divided into three parts. The southernmost, in the Himalayan foothills, is known as Solu. Its western boundary descends southward from Numbur mountain for about 36 kilometres (22.5 miles) along the Likhu Khola river and extends eastward for about 50 kilometres (31.25 miles). The middle region is called Pharak, consisting of a 20-kilometre (12.5-mile) forested stretch of the Dudh Kosi river valley, from

Khumbu, the Sherpas' name for the most northerly patch of their homeland, is hardly more than three long and slender valleys that descend from the heights along the Tibetan border. The upper reaches of each of these valleys are covered by daunting glaciers, born from the snowy mantles of some of the world's tallest mountains. With names like Khumbilia, Cho Oyu, Gyachung Kang, Pumori, Taboche, Kangtaiga, Lhotse, Nuptse and Chomolungma, the Mother Goddess of Earth, they form a backdrop for each Khumbu village so majestic and spectacular that no Sherpa can forget how humble Man is compared to the forces of nature that made this landscape a place of unique beauty and yet unforgiving harshness. Thanksgiving is a daily occurrence here. These extraordinary people, so full of life, are just happy to be able to survive in their hardy environment, where with each season there is less wood to cut, more soil erosion and less fodder for their yaks. They know how to count their blessings and to make the most of what they've got. They bless each cup of *chang*, a milky home-brewed rice or barley beer, and *raksi*, their firewater distilled from potatoes, before drinking it – and they drink a lot. The ritual is simple and touching: they daub the rim of their glass once with flour, then flick a droplet of the nectar into the air for the gods to sample, at the same time thanking them with a mumbled *mantra* (prayer) for having revealed to mere mortals the secrets of fermentation and distillation. Sherpas consider *chang* and *raksi* as gifts of the gods, their very special gods, who fawn over and safeguard them.

The Sherpa clans that first filtered into Khumbu's hidden valleys almost five hundred years ago came from eastern

---

the village of Monjo southward to the Khari Khola river, a tributary of the Dudh Kosi. The most northerly, bordering on Tibet, is Khumbu. Of the three, it was settled the earliest, the first Sherpas arriving there in the mid-16th century. Soon after, some Sherpa clans pushed southward and settled in the more hospitable and fertile Solu. The three regions together are referred to as Solu-Khumbu. In all, about 25,000 Sherpas live in Solu-Khumbu. Sherpa colonies also exist in Kathmandu and the Indian mountain district of Darjeeling.

Tibet. They crossed the Nangpa Pass over the Great Himalaya divide to find shelter in Khumbu from the wars that were then ravaging central Asia. As a result, Khumbu's culture is more Tibetan than Nepali. Until recently the inhabitants, Buddhist in their beliefs, looked more to the Tibetan capital of Lhasa, the spiritual centre of Tantric Buddhism, than to Kathmandu. Because of this, Khumbu is rather more detached from the rest of Nepal than the King and his ministers, the country's Hindu hierarchy, care to admit or feel comfortable with and consequently the government is making a serious effort to 'Nepalize' the Sherpas by better integrating them into the country's social and economic life.

Before meeting with Gyalzen next morning, I had some business to complete. Principally this consisted of a visit to Dr Hemanta Mishra, head of the King Mahendra Trust for Nature Conservation. I had met Hemanta Mishra in London the year before and told him about my project. At first he had been sceptical but now he was warming to the idea that a yeti might exist, in which case it would become his responsibility, as Nepal's leading environmentalist, to protect it.

Almost single-handedly Dr Mishra had fought to give Nepal one of the world's most comprehensive conservationist policies. While at the Department of National Parks and Wildlife Conservation, he had structured the programmes and legislation for the country's sixteen national parks and wildlife reserves. He had a powerful friend and ally in Prince Gyanendra Bir Bikram Shah, the King's brother, who is chairman of the King Mahendra Trust.

Hemanta provided me with a letter of introduction to Nima Wanchu, the warden of Sagarmatha National Park, which includes most of the Khumbu region. He invited me to a party for his staff later in the week to celebrate his having been awarded the Paul Getty Prize for nature conservation. I would meet at Hemanta's house Dr Eric Dinerstein, the Smithsonian Institution biologist working on the Greater One-horned Rhinoceros Project. The one-horned rhino (*Rhinoceros unicornis*) is a survivor from the Pleistocene period. A small herd lives in Chitwan valley, a marshy belt of the Terai that stretches along the foothills of the Nepal Himalayas. Hemanta, a doctor of animal biology, regarded these animals as one of our last living links with prehistory, unless of course I could produce for him the yeti and prove that it is a relict primate. Only 1,500 of the unicorned rhinoceratids survive as they were almost hunted into extinction for their horn, which is deemed both magical and medicinal, selling in the pharmacies of Asia for $30,000 a kilogram. Dinerstein, in his mid-thirties and fluent in Nepali, was quietly fascinating about his work, a sort of dedicated Gentle American who is a most effective ambassador.

Before leaving Kathmandu, I also wanted to see one of Hemanta's protégés, Bijaya Kattel, who is Nepal's senior wildlife officer. A few months before, Kattel had begun a study of the Himalayan musk deer. This had taken him to a village in Khumbu near where I would be located. But Ghanashyam had told me soon after my arrival that Kattel was back at his home in the suburbs of Kathmandu, recovering from a near fatal bout of typhoid fever. When we spoke on the telephone, Bijaya said he was hoping to return to Khumbu soon, but that his convalescence had been delayed for an additional few weeks. I asked why and with a laugh he explained that, feeling better, he had hopped out of bed, stumbled and broken a foot. In addition to becoming a good friend, Bijaya filled the role of sounding board and was an excellent counsellor.

Next morning Gyalzen was waiting in front of the Yeti Mountaineering and Trekking office in Ramshah Path, not far from the Russian cultural centre. With him was Pasang Tshering, a timid 13-year-old who spoke no English. Pasang Tshering had only been to Kathmandu once before in his life, and that was as a baby. He seemed positively overwhelmed by the city, its noises and the rush of motor traffic. He stayed close to his father, was obviously eager to be helpful and did whatever he was told with a fervour that brought him as close as he could come to perfection.

It was our first encounter since the spring of 1986 when I had hired Gyalzen to be my guide on a reconnaissance of the upper Gokyo valley (one of the three valleys of Khumbu) and Cho Oyu, the peak at the valley's northern limit. We had met on a wonderfully warm March day in Namche Bazar, a dusty trading centre that nestles in a deforested bowl surrounded by mountains, each the domain of a much adulated Sherpa god. Gyalzen had been highly recommended by his brother-in-law Lobsang Tshering, who lives in Bodnath, the Sherpa village on the outskirts of Kathmandu. As Gyalzen had been on the treacherous slopes of Cho Oyu before, serving as a high-altitude porter for a foreign climbing expedition, I decided to take him with me on the reconnaissance. His high-altitude portering made him one of the local elite known as 'tigers of the snow'. His climbing experience, Lobsang said, was considerable.

During that 1986 reconnaissance, which lasted four weeks, we had gotten on well together, for Gyalzen was not only a grass-roots philosopher and religious scholar but a natural born optimist. We had climbed from the yak herders' hamlet of Gokyo through the snows on the far side of Macherma Peak to Renjo Pass, at 5,417 metres (17,768 ft) on a barren route said to be used by the yeti in its travels from one valley to another. Before returning to Namche Bazar, we crossed the gravelly tongue of Ngozumba Glacier, climbed Tshola Pass, and descended by Duglha on the east side of the pass to Pheriche, en route for Gyalzen's home at the tiny Sherpa settlement of Pangpoche, another half-day's walk to the south. Gyalzen had invited me to stay for a few days at his home, a two-storey stone and mud dwelling with wooden beams and flooring, no running water or heating, and a wood shingle roof. There I met Kanche, his wife, and their four children, who treated me to the kind of hospitality that only people who live by the gifts of nature, possessing little material wealth, can bestow upon a traveller. It was wholehearted and warm, something so utterly touching that I knew I had to return.

Throughout our 1986 trip together, we had seen no wildlife but had followed the footprints of a wolf to the crest of Renjo Pass and uncovered the nest of a snow leopard above Tshola Tsho. It was then that Gyalzen told me that Sherpas believe that wherever the snow leopard prowls, the yeti also is present. When later I wrote to cryptozoologist Dmitri Yurevich Bayanov in Moscow about this, he replied that the great Russian expert on relict hominids, the late Professor Boris Porshnev, had heard the same thing from Kirghiz tribesmen in the Pamir mountains of central Asia. Prof. Porshnev investigated further and came to the conclusion that the yeti scavenged off the snow leopard's kills. To me this seemed entirely plausible. Prof. Porschnev further believed that the yeti was in all probability a relict hominid. I had not yet learned that this type of scavenging for food was common practice among prehistoric hominids such as *Homo Erectus*. Once I uncovered this behavioural similarity, the yeti enigma assumed new dimensions for me.

Now that I had met up with Gyalzen again, we went into Bharat Parajuli's office and made a shopping list — rope, kerosene lantern, down jacket and boots for Gyalzen, a good pair of shoes and a rucksack for Pasang Tshering, some canvas sheeting, batteries, film canisters and food supplies — for our return to Pangpoche. Gyalzen said he would see about hiring sixteen porters to carry our material from Jiri to Namche Bazar where another brother-in-law, Pasang Sherpa, would be waiting with a string of yaks to take us the rest of the way up the Imja valley, past the monastery at Tengboche, to the terraces and potato patches of Pangpoche.

While we waited for the expedition drums to arrive — they had been delayed at Delhi during the ten-day festival of Dasain — we had time to explore the city and its valley. All work grinds to a halt during Dasain, a popular festival because it celebrates the victory of the much-loved Rama, seventh incarnation of the god Vishnu, over Ravana, the most dangerous and powerful demon king ever to have existed in the annals of Hinduism. The copper-eyed Ravana kidnapped Rama's wife, Sita. The world stood still while the heart-broken Rama meditated for ten days, deciding what to do. Ravana was no ordinary match. He had ten heads, teeth shaped like the crescent of a young moon, and twenty arms.

He was the leader of the rakshasa warriors, many of them gorilla-like and hideous to behold, others dwarf-like or one-eyed, with monstrous bellies and crooked legs, still others with animal heads, all told a horrible host. Finally Rama enlisted the help of the fierce and powerful Kali, the black earth-goddess who wears a necklace of skulls and whose rites involve sacrificial killings that once were human offerings but now are associated with the slaughtering of animals. In an epic battle Rama and his many supporters defeated the rakshasa generals. Then Rama took his revenge, slaying Ravana and rescuing his beloved Sita. The ten days of Dasain are riotous, with much revelry, slaughtering of animals and colourful processions through the city.

Kathmandu entered the twentieth century with some delay but has been working hard to catch up ever since. So successful has it been in this race that now it is threatened with asphyxiation by the internal combustion engine and other modern inventions. Most, though by no means all, of the streets are lit at night, a considerable feat, since the city only received around-the-clock electricity in 1982. Although its streets are congested, I marvelled over the illogical fluidity of the traffic in spite of the almost total absence of traffic lights. Then one day as Gyalzen and I were walking up Durbar Marg we noticed workmen installing a set of tricolor signals. Soon traffic lights had sprung up all over Kathmandu, in preparation, I learned, for the inaugural summit meeting of the South Asian Agency for Regional Cooperation (SAARC), headquartered in a new building on Kantipath. Alas, I viewed this as further evidence of an international conspiracy to force the storybook kingdom to catch up with the rest of the world. But Kathmandu still has no sewage system, and garbage disposal remains a problem. Nor does anybody know for sure how many people live in the city: it is estimated to be somewhere around 300,000, though it could be 500,000. But no one was unduly worried by this: life continues as it always has and the Kathmandians, if we can call them such, are colourful, industrious and appear to be happy.

Some of the tourists we met in our wanderings did seem to grumble a lot. They complained that the tap water was almost lethal, which is a fact, though the native Star Beer is good. Watch out for giardia, they warned, looking sheepishly at Gyalzen who was at most times circumspect and said little. At first I thought by giardia they meant that somebody's pet gerbils had run amok. But no, giardia are unfriendly water-borne amoeba that hatch in your intestinal tract and can make life very unpleasant, though I am told that an old-fashioned drug called metronidazole (marketed in some countries as Flageole) provides a painless, effective and simple cure.

Then also I often heard it repeated that the King of Nepal is among the twenty richest paladins of this planet, while the country he governs as a divine though constitutional monarch is listed by the United Nations as among the world's dozen poorest. As frankly I doubted that anyone in Nepal could possess such fabulous wealth, I consulted *Fortune*'s list of the world's richest — the 129 men and women from many lands whose assets top $1 billion. Rest assured, King Birendra Bir Bikram Shah is not counted among them, the richest man in the world being the Sultan of Brunei, who comes far ahead of the runners-up, King Fahd of Saudi Arabia, the Mars family of Chicago, and Queen Elizabeth II in that order.

In spite of its many failings, including the added confusion that the traffic lights brought to its congested streets, modern Kathmandu retains a colonial quaintness, a whiff of the British Raj. The King and Queen live in an art-déco pagoda-style palace surrounded by red and yellow poinsettia trees at the top of Durbar Marg. Here and there are hidden treasures of neo-classical architecture, for the most part the dilapidated palaces of the Ranas, and in the back streets of old Kathmandu as in the surrounding countryside are gems of traditional Newar architecture, also, alas, a little dilapidated, a little forlorn, like an Oriental fairy tale that has become faded with the passage of time and battered by an earthquake now and then.

Of course what really distinguishes Kathmandu from other cities are its people: Newars mainly, but also a mixture of other ethnicities with their own dialects and cultures, more than thirty of them from throughout the kingdom, an amalgam of north and south, of Tibeto-Burman and Indo-Aryan — of Tamangs and Rais, Sherpas and Thakalis, Gurungs and Limbus, Magars and Bhotias, and, increasingly, a mix of illegal Indian immigrants.

For most anthropologists, the Newars are the original inhabitants of the valley, though they have been overrun and dominated by the lords of so many successive invasions that nobody is quite sure when and from where they came. They are said to have originally been Tibeto-Burman and Buddhist, but over the centuries they have absorbed Indo-Aryan characteristics and are now slightly built, graceful, and mainly Hindu. They are superb farmers, craftsmen, traders and builders. Their economy, like their culture, is a unique blend of the primitive and the sophisticated.

They still till their fields with Stone Age mattocks, yet their irrigation systems are complex and efficiently run as village cooperatives. They harvest three crops a year. Their homes are attractive and their temples ornate: brickwork walls shaped around carved wooden frames, capped with pagoda roofs. Their villages resemble what I imagine medieval German or Swiss villages might have looked like half a millennium ago. They are usually built around holy sites, along the spines of ridges, to save the more arable land for productive cultivation.

Kathmandu Valley is a land of legend. Buddhists believe that the valley was once a turquoise sea upon which floated a single lotus flower. This blossom was said to be the manifestation of the primordial Buddha known as Swayambhu. From China the great warlord Manjurshri came to worship the lotus blossom. To get closer to it he smote Chobar mountain, to the south of Kathmandu, with a flaming sword, parting it so that the water drained from the lake. Where the lotus blossom settled, he founded a Buddhist temple, Swayambhunath, which remains one of the most venerated shrines of Buddhism.

The Hindus, of course, are convinced it did not happen this way. The valley, they believe, was drained by Lord Krishna who hurled a thunderbolt at Chobar mountain to slice it in two.

Whether created by flaming sword or thunderbolt, the Chobar gorge exists, so narrow that neither man nor beast can pass between its walls. It can be crossed, however, by a suspended iron footbridge manufactured in Scotland, carried into the valley in sections by bandy-legged Nepali porters and erected in 1904.

One day before the expedition drums arrived, Gyalzen and I went to the Narayan temple on a hilltop at Changu, 12 kilometres (7.5 miles) east of Kathmandu. For me this temple is one of mankind's least known art treasures. Said to be the oldest pagoda-style building in the world, it dates from the fourth century, but was rebuilt after a fire in 1702, and damaged again by the earthquake of 1934. We walked to it along the tiled pathway of tumble-down Changu village, which half a century later still shows signs of the devastating earthquake.

The temple is dedicated to Vishnu, one of the Hindu holy trinity. Outside its main entrance pilgrims are greeted by a pair of stone elephants and the image of Garuda, the mythical bird that serves as Vishnu's vehicle. Some of the carvings in the brick-paved courtyard are said to be from the fifth, sixth and ninth centuries. But I found no *ban manus*, or forest man, carved in stone and was disappointed that the Hindu wildman was not represented in Narayan's pantheon of carved images. This reminded me that the yeti is very much a Buddhist animal and is hardly represented at all in Hindu mythology.

Changu's hilltop offered a sweeping vista, with the snow-capped giants to the north tugging at us to begin our caravan march into Khumbu. To the east we found a path that descended to Changu's rice fields and the flat plain that stretches like a carpet to the walls of Bhaktapur, the eighteenth-century capital of the Malla kings, 4 kilometres (2.5 miles) south of us. At the foot of the terraced hillside, we encountered the local guild of brickmakers, practising their craft as they have done for centuries. We watched in fascination as men and women extracted earth from the field they were exploiting. They mixed the earth with water and patted the resulting mud into simple wooden moulds with the guild's trademark embossed on them. Once the bricks were tipped out of their moulds, they stacked them to dry in the sun. The brickmakers had rented the field from local farmers and after extracting the agreed-upon quantity of mud, they would move to another field and start the same process over again. Some of their bricks, the more expensive variety, were fire-baked in primitive kilns.

From the fields beneath Changu we made our way to the medieval village of Thimi, 3 kilometres (1.9 miles) west of Bhaktapur. Thimi's name is derived from *chhemi*, meaning 'capable people'. Today its citizens are renowned as Nepal's most capable potters. Men and boys shape smooth reddish clay on ancient wheels, producing cooking pots, water jugs and other vessels. Their wares are then fired in the streets under huge piles of straw. Porters come from all over the valley to carry Thimi's pottery to distant markets. Before leaving the village their enterprise is blessed at the sixteenth-century temple dedicated to the female goddess Balkumari, the *shakti* of Bhairav, god of fertility, much as wool merchants might have appealed to a patron saint before setting out from Bruges or Florence in the Middle Ages.

Returning to Kathmandu in the late afternoon, the delights of the 'modern' city awaited us. We drove past the Tundikel parade ground, Kathmandu's central park, where Gurkha soldiers practise their drill, and the Rani Pokhri, a lake where the Ranas once dispensed justice by dunking suspects in the brackish water. Here we turned left into New Road, the main shopping thoroughfare. We still had some bricks of Tibetan tea and candles to buy. A short walk from New Road is the templescape of Durbar Square and the marvels of Freak Street. This narrow alleyway lined with book shops, cheap eateries and curio sellers was once the lair of hippies and still is the haunt of dope sellers and currency blackmarketeers.

More inviting was a walk through the markets of old Kathmandu, alive with colours and scents and the wares of several continents. We marvelled at the ad hoc shrines and temples, some only niches in the wall, or a phallic carving in a sunken pit by the roadside. In one backstreet, we found a stump of wood, not more than 80 centimetres high, set in its own quadrangle by the side of the roadway. Dedicated to Vaisha Dev, the god of toothaches, the stump was studded with a thousand nails, placed there by toothache sufferers. Each nail represented a request to the tooth god to guide the hand of the traditional dentist who would soon treat the toothache, imploring that the cure be as painless as possible. My teeth were not troubling me yet, though they would later, and so I passed by the tooth god's temple without hammering a nail into the stump.

Even though I was constantly worried about our air cargo of expedition drums and skis, the week spent with Gyalzen exploring Kathmandu and its valley, seems in retrospect a restful, enchanting interlude before the strain of our travels in Khumbu. But it was not to last. Returning to the guest house on the night of 15 October a message was waiting that the drums had arrived. We went to the airport to collect them next morning and it took most of the day to clear customs. Surprisingly, everything was intact. The only casualty was the ski bag. Someone had cut off its straps.

Once cleared through customs we assembled the drums in the front garden of the guest house and, with the gear I had already stockpiled there from previous visits, repacked equipment, clothing and food into more portable loads, according to the priorities of the trek to Pangpoche. Pascal Marlinge, a 34-year-old Frenchman helping Parajuli and Paudel organize French trekking groups, was staying at the guest house and he, too, pitched in, reorganizing our pharmacy in a separate tin trunk that we purchased in the market-place. John Thornicroft, the Channel 4 producer, and his presenter, entertainer Mike Harding, had dropped off our tents and additionally left us with 35 kilograms of filmstock which they wanted transported to Namche Bazar. They said they would reimburse us for the porterage, they having flown in the meantime to Jomoson to film in the Annapurna Sanctuary. Their primary interest, though, remained the garbage discarded by trekkers and climbers on the 'Kleenex Trail' to Everest Base Camp, which was something of an ecological disaster zone.

Now that we had everything assembled, we planned to leave Kathmandu at dawn on Monday, 19 October. Before returning to his brother-in-law Lobsang's house at Bodnath that Friday evening, Gyalzen said he would hire porters, cook and kitchen boy during the weekend. And so I spent my last two days in civilization quietly relaxing, going over inventories and sending off my last mail to Europe, to my wife in Canada and our three daughters. The weather was close, but not as heavy as the mood in New York. Since Wednesday, Wall Street had been hit by a wave of panic selling and the Dow Jones Industrial Average closed that Friday at 2,246, off almost 300 points for the week. Friday's volume was 350 million shares, but the play in the marketplace, as I later appreciated, was only beginning to warm up. 'It's panic city out there,' one commentator was quoted as saying in the *International Herald Tribune* I had bought that day in Thamel. The fact that New York had become 'panic city', and other major stock markets were trembling in anticipation of darker events to come was something that entirely escaped the attention of the Nepalis.

Kathmandu, of course, is far removed from Wall Street. But even this out-of-the-way Asian capital has its own stock exchange, located in a two-storey villa at Dilli Bazar, on the road to Battisputali. Only twenty-five companies are listed for trading on the blackboard at the back of what was once the villa's dining-room. Trading sessions last for two hours each morning, five days a week, public and religious holidays such as the ten-day Dasain festival excluded. They are pretty sleepy sessions at the best of times and the trading volume is counted in terms of a few hundred shares, not the few hundred millions in New York. The Kathmandu Stock

Exchange was one market completely unfazed by the black hole that almost swallowed other stock markets when trading resumed on Monday. Dilli Bazar seemed pretty much detached from the rest of the galaxy, existing in a universe of its own — a very small universe at that. Of more concern to Kathmandians that Friday was another kind of storm, this one building up in the Bay of Bengal, 850 kilometres (530 miles) to the south-east.

Chapter 2

The Storm

Kathmandu is on the same latitude as the Canary Islands and Tampa in Florida. As it is situated at an altitude of 1,300 metres (4,265 ft), the climate is moderate, never too warm and rarely very cold. It receives most of its rainfall during the summer monsoon period, from mid-June to mid-September. Autumn and winter precipitation is unusual and at Kathmandu's altitude it never snows. Saroz and I were mildly surprised, therefore, when it started to rain on Sunday evening as we returned from dinner in Thamel. Though only a few drops, not enough to dampen the dust in the streets, it told us that the cyclone approaching the mouths of the Ganges was pushing bad weather towards the Himalayas.

In the emerald valley the cloud ceiling remained high, but air was close and the night sky leaden. Later, as I lit the herbal anti-mosquito coil beside my bed, silent sheet lightning rolled across the valley. I was not unduly concerned, however, as I expected the storm would pass in the night. Its full force only broke around 3 a.m.: pounding rain, with bursts of fork lightning that lit the room for seconds at a time, followed by claps of thunder that literally shook the bed. By 4 a.m. I could no longer sleep. The storm had developed into an unbroken downpour. I got up and started repacking my rucksack. Good morning, Black Monday, 19 October 1987.

That weekend a Dutch couple in their early twenties had arrived at the guest house and shyly announced that they intended to go trekking. But as they had never been in the mountains before, they wanted to join a group. They made their intentions known with the halting discretion of two climbers planning an assault on Everest but who feared being ridiculed for their lack of basic climbing skills. I said I would take them, if they wanted, as far as the Sherpa market village of Namche Bazar for a fee of $40 a day, all included, to help cover my costs. They agreed, rented sleeping bags, and hurried off to purchase their trekking permits from the immigration office.

To get from Kathmandu to Gyalzen's village of Pangpoche, our operations base for the winter, we had to travel 344 kilometres (215 miles) through mountainous country, crossing four passes and nine rivers, two of them without proper bridges. As only 180 kilometres (112 miles) were on paved roads and, because of a four-day Hindu holiday during which few porters would agree to work, it would take us twelve days. Even the paved roads did not offer secure conditions. In one place, before the central hill village of Charikot, we had to portage around a landslide. Then on the 30 kilometres (19 miles) of semi-paved track from Charikot to the roadhead, whole sections were in danger of being washed away by the Black Monday storm.

At Jiri we would leave the twentieth century and walk 154 kilometres (96.25 miles) on dirt paths to Namche Bazar, where Gyalzen's brother-in-law Pasang would be waiting. After buying more supplies at Namche's weekly market, we would set out on the last 14 kilometres (9 miles) of steep trails to Pangpoche. We intended using yaks to transport our baggage over this stretch, above the Dudh Kosi river to the Buddhist monastery of Tengboche. From Tengboche the trail goes through a birch and rhododendron forest, crossing the Imja River, a tributary of the Dudh Kosi, on a suspended footbridge, then climbs under slabs of grey granite to the first potato fields of Pangpoche.

Along the way we would stop at tiny tea houses to sample the home-brewed *chang*, a milky beer-like drink, pass beside hand-carved prayer stones, rows of water-powered prayer wheels, and flapping *tarchens* — long white Buddhist prayer flags. The final approach to Pangpoche is guarded by a votive *chorten*[1] and a free-standing gate to keep out evil spirits. Like all Khumbu villages, it has never known a wheeled vehicle, not even a wheel barrow, the only means of transport being the backs of humans or yaks. It is in another world, that, other than the scattering of non-biodegradable trash, has hardly been touched by the passage of time.

Parajuli's trekking agency had chartered a bus to take us as far as the Charikot landslide. The bus rental was 8,000 rupees ($320), well worth it.[2] On the far side of the landslide we would have to hire another bus to take us the rest of the way to the end of the 'road', though it left us still 154 kilometres (96.25 miles) from our destination.

The road is so tortuous, with never-ending switchbacks and stretches where only one vehicle can pass, that under normal conditions the 180 kilometres from Kathmandu to Jiri take a full six hours to travel. The bus, with Gyalzen, Pasang Tshering, cook, kitchen boy and sixteen porters, should have arrived at dawn. But Ghanashyam Paudel telephoned from the agency at 6:30 a.m. to say that Gyalzen had not turned up with the porters. The rain seemed heavier than ever and I could imagine most porters would prefer to stay home on such a day. An hour later, Ghanashyam called again to say that the bus was on its way.

Alas, it was a big grey Tata capable of seating fifty-two passengers, too big to manoeuvre into the alleyway that led to the guest house, and so it had to park in Battisputali's

---

[1]A *chorten* is similar to a *stupa*. It is a dome-shaped Buddhist monument of fixed architectural design that represents one of the phases of Lord Buddha's enlightenment. It generally encloses sacred texts or liturgical objects intended to ward off malevolent spirits. The eyes often painted on a chorten represent the all-knowing eyes of Buddha. Chortens sit on five-step pedestals, each step representing one of the basic elements of life: earth, water, fire, air and the cosmic universe.

[2]Throughout, I have used an exchange rate of 25 rupees to the US dollar.

main street. This meant that we had to carry our gear out to it through ankle-deep puddles. I have never experienced such driving rain. It was now 8 a.m. and it had been falling in unbroken sheets for four hours. Parajuli, his eldest son, and Pascal Marlinge had come to bid us safe journey and finally, soaking wet, we climbed aboard. Our driver started the motor, rammed the gear stick into an appropriate slot and off we shuddered like a steamer pulling out of port.

The metal floor plates moved and the door kept sliding open to let in more rain till the number-one assistant stopped banging out his navigational instructions to the driver on the side paneling and jammed the door shut with a metal strut. Nor would all the windows close, and it seemed, suddenly, miserably cold. I was soaked and steaming, but reflected philosophically that the rain could not continue for long. My reasoning was simply that the heavens were unable to hold so much water.

The Tata works in India churn out thousands of these buses each year, based on an old Mercedes design, with tough and basic engineering to endure the subcontinent's wondrous roads. Our beat-up model had an eight-speed gearbox similar in many respects to a coffee grinder. Moreover, our windshield wipers didn't work, the road was slippery, and I had my doubts about the brakes.

We passed the airport and turned east onto the Arniko Highway. I couldn't relax, so I walked down the aisle taking stock: all eleven drums were safely on board, as well as our rucksacks, ski bag, portable pharmacy and five *dokos* — about 550 kilograms in all. A doko is a wedge-shaped wicker basket used by the Nepalis to carry loads. They attach a tumpline to it, which they slip over their head to support the basket on their backs, and in this manner they can carry more than 30 kilograms (66 pounds) over long distances and difficult terrain. Most of Nepal's inter-village trade is transported in these baskets. We also had a cook, kitchen boy and, I expected, sixteen porters hired by Gyalzen. With the three of us and the Dutch couple that should have made twenty-three in all. I had noticed by then that our charter was staffed by a driver and three assistants who sat in a separate compartment up front. The Dutch couple — Marjolein Schade and Jaap Zondervan — sat immediately behind the passenger door which, in spite of the metal strut, refused to stay shut. Jaap had propped his foot against it.

I looked around for the porters. Aside from Gyalzen and Pasang Tshering, I counted only two other persons.

'How many porters do we have?' I asked Gyalzen.

'Just no porters. We have one kitchen and one Sherpa,' he replied. In his halting English he explained that the candidates he had rounded up on the weekend failed to show that morning and that he had not been able to find replacements.

'What are we going to do?' I asked.

'Jiri we find plenty,' he assured me.

'And for the landslide?'

'Many boys there,' he said. 'We hire some.'

Pasang Tshering, for certain, was pleased to be on his way home. He had been in Kathmandu with his father for nearly a month and while he enjoyed seeing his two cousins in Bodnath, who were going to a nearby school and already were quite proficient in English, he was eager to see his mother again and resume his duties as the eldest son, very much the little master of the household who looked after Mingma Yanchin, his sister, one year younger than himself, Pemba Kunga, his nine-year-old brother, and Tsheden, the infant tyrant. He had his adventures with which to regale them and his new possessions such as the blue plastic baseball cap marked *Captain* and the oversized blue wind jacket he had on. His window was wide open and he was peering out at the rain-soaked countryside. The fields were green, but the rain had flattened a good half of the rice crop.

The Arniko Highway was built by the Chinese in the mid-1960s and it runs to Kodari on the Tibetan border, 110 kilometres (68 miles) to the north-east. Arniko was the name of a Newari architect who in the thirteenth century was invited to Lhasa to build fabulous monuments for the then-ruling Dalai Lama. Word of Arniko's genius was soon carried to the Ming emperor of China who asked the Newari builder to his court and appointed him 'comptroller of imperial manufactures'. The concept of multi-tiered pagoda temples, which became a trademark of Eastern architecture, was brought to China by Arniko and because of this cultural legacy the Chinese authorities under Chairman Mao regarded him as the master craftsman of Sino-Nepali friendship; hence they named the road, a major engineering feat, after him.

If you placed Nepal on a map of the United States it would stretch end-to-end from Nashville, Tennessee, to Dallas, Texas. But distances in Nepal, where most of the nation's trade is carried on the backs of porters, are not the same as the rest of the world. The country only has 2,080 kilometres (1300 miles) of motorable roads and 99 kilometres (62 miles) of railroad track. Although a section of the Pan-Asian Highway crosses southern Nepal, to traverse the country's full length of 880 kilometres (550 miles) is an adventure that can take several weeks. We were only attempting a fifth of that distance and it would last for a day-long eternity: twelve of the wettest hours I have ever known.

We reached the toll gate at Banepa, 19 kilometres (12 miles) east of Kathmandu, at 9 a.m. and found the roadway blocked by six bogged-down buses. We paid the toll and managed to squeeze by them, continuing eastward into the grey, beating rain. Everywhere sheets of reddish-coloured water ran off the hillsides. At one point we were stopped by a herd of black and white goats standing in the middle of the road – maybe two hundred of them – having walked all the way from Tibet. They looked homesick and no doubt sensed they were approaching the end of their journey, where a Bergen-Belsen for innocent capricorns awaited them.

A little after 10 a.m. we crossed the Sun Kosi. *Kosi* in Nepali means river, and *Sun* is their word for gold. The river sweeps down from Tibet and eventually joins the Ganges, its waters flowing into the Bay of Bengal. Normally the Sun Kosi is clear and sparkles like gold, but today it was an angry brown, laden with silt which the hammering rain had washed off the denuded hills. Tiny mountain streams had swollen into torrents that ran from the terraces of yellow rape seed into green rice paddies and pastures scorched brown by the post-monsoon heat. The water run-off was unrelenting and blood-red in colour as if the hillsides were bleeding.

In one village our Tata stopped and the driver talked with the driver of another bus coming from the north. The south-bound driver warned that the road was becoming precarious and if the rain continued much longer would soon be impassable. There was not much traffic in any event as the Chinese had closed the frontier with Tibet after rioting had begun in Lhasa three weeks before. The riots, every Buddhist in the Kathmandu suburb of Bodnath was aware, had left scores dead, hundreds injured and many thousands more under strong-armed arrest. Rumours had been rife in Bodnath before we left that Chinese troops were about to storm the Johkang, holiest of Tibetan monasteries, considered a hotbed of rebellion by the Chinese authorities.

When Gyalzen saw me pull out my soggy newspaper with a picture of protesting monks on the front page he asked me to read him the news from Lhasa. He was worried that the Johkang would be damaged and the monks slaughtered. On the same front page, figuring much more prominently, was news about the panic on Wall Street. But Gyalzen didn't know about this temple of capitalism and the American Dream was not even a glimmer in his mind. Thank god, or rather thank Buddha, that there are still people like Gyalzen who refuse to believe — nay, have never considered — that Wall Street is the navel of the world.

Our driver banged the Tata back into gear, jerking us forward. The noise of the rain hammering on the roof sounded like hail. We crossed a series of old washouts only partly repaired since the last monsoon, our outer wheels centimetres from the crumbling drop-off to the Bhote Kosi, a hundred metres below. The river was carrying tree trunks and other debris while the countryside seemed ever more raw under the beating rain.

By 11 a.m. we reached Bahrabise, 60 kilometres (37.5 miles) east of the Banepa toll gate. At Bahrabise the road divides, north to the Tibetan frontier or eastward across the Bhote Kosi to Jiri. There are in fact many Bhote rivers in Nepal as *Bhote* means 'from the north' and most Nepali rivers flow southward from the Great Himalaya divide. In Bahrabise the rampaging Bhote Kosi had already submerged parts of the road, to the satisfaction of a sow and her piglets who were wallowing in the pools of run-off water and mud, having just the most pleasant time imaginable. Four other buses were parked by the roadside and the two local cafés — wooden shacks, really — were crowded. We were now 70 kilometres (44 miles) from Kathmandu and had another 110 kilometres (68 miles) to go before reaching Jiri.

For lunch in our roadside shack a soot-covered woman in a long black shift served us *dal bhat*, the standard Nepali dish of rice, lentils and curried vegetables. Adopting local custom, we dipped in with our fingers, washing the spicy food down with glasses of treacly milk tea that the Nepalis call *dudh-chiya*. Concerned that the road might soon be washed out, the driver only allowed us twenty minutes for lunch. I paid the bill, which came to $4 for seven of us, and we got under-way again, crossing the Bhote Kosi on a girder bridge wide enough for one vehicle. The river's angry waters were as grey as our bus and leapt at the bridge's underside with menacing force. Everyone aboard was relieved when we rolled off it and began negotiating the first series of switchbacks on the Jiri Road, built by Swiss engineers twenty years before. The rain was like a curtain, never letting up. I kept telling myself that it could not continue like this, but I was mistaken.

We climbed for an hour into the Mahabharat mountains, a middle range that separates the jungle plains of Terai from the Himalayas. Rectangular in shape, tilting to the south-east, Nepal is between 145 kilometres (90.6 miles) and 190 kilometres (120 miles) wide. In that short interval, the terrain changes from tropical jungle to arctic tundra, from plains that are at sea level to peaks of more than 8,000 metres (26,000 ft.). The Tata laboured through its gears, having to stop on the hairpins, back down again and have another try. Finally around 2 p.m. we reached the landslide a few kilometres before Charikot, a village of maybe 500 persons that had a police checkpoint and was the administrative centre for Dolakha district in east-central Janakpur province.

The landslide occurred during the summer monsoon, when moisture-laden air from the Indian Ocean is sucked towards the seasonal south Asian low-pressure system suspended over the Tibetan plateau. As the warm air ascends the slopes of the Himalayas it is cooled, clouds form and heavy rain results. Since Man has known about such things, the monsoon re-occurs with clockwork dependability, providing the subcontinent with most of its annual precipitation. Several months would be needed before the government could assemble the necessary engineers and equipment to clear the Charikot slide. It was one of hundreds caused by the rains each monsoon season. And with every monsoon the landslides become more numerous and above all more murderous as the on-going deforestation leaves the exposed slopes more prone to devastation.

We parked in a big bend of the road and unloaded everything into the rain. We stored as much as we could under the green nylon sheet we had bought in Kathmandu. Gyalzen began looking for porters to convey our gear to the other side, twenty minutes away. As luck would have it, a large French group led by Jean-Franck Charlet of Chamonix, intending to climb Baruntse, had pulled in ahead of us and snapped up all available help. As it was no use standing around in the rain, I loaded one of the lighter drums onto my rucksack and set off down the road to a slippery path that descended to the swollen Charnawati river, crossed it on the few greasy planks that served as a bridge, then climbed a muddy trail to rejoin the roadway on the far side of the river. It must have been an unpleasant passage at the best of times, but it was pure hell in the rain.

Pasang Tshering was left to guard the material stockpiled under the nylon sheet. On the Charikot side Marjolein and Jaap shepherded our goods under some black plastic bags extracted from one of the barrels. They were looking anxious and uncertain, as if regretting they had become involved in such a mad venture. The rest of us — Gyalzen, myself, Bir Bahadur, the cook, and Nawa, the kitchen boy — made three trips each, with Gyalzen hiring a few schoolboys to help on the last one. We were, all of us, wet to the core, looking half-drowned, and the rain kept pelting down.

Two buses were parked on the far side of the landslide. Gyalzen began negotiating with their drivers. Meanwhile the scheduled bus had arrived from Kathmandu and was disgorging its passengers into the rain, so that with the French group and ourselves there was now a heavy demand for transport to Jiri. The drivers, of course, knew how to exploit this situation. The French commandeered one bus. We chartered the other for 5,000 rupees ($200), and began loading our gear.

It appeared that the regular passengers would have to wait, under the dripping canvas of a tea shop that had been temporarily installed by the roadside, for a third bus to arrive from Charikot. But as our conductor was not one to miss an economic opportunity, he counted the number of empty seats and took on board about thirty of the stranded travellers, charging them the normal fare even though the bus was supposedly fully chartered. Looking at it from another angle, we were charged the usual fare plus 8 rupees (32 cents) per kilogram of freight to keep our baggage out of the rain and securely stowed in the aisle and seats at the back of the bus. In any event the conductor slipped an extra 1,000 rupees ($40) into his pocket, more than the average monthly wage, which he would split with the driver.

The road from Charikot to Jiri, a distance of 30 kilometres (18 miles), seemed long. It was soon dark and I was shivering, but could not squeeze my way to the back of the bus to dig out a jacket and dry shirt from my rucksack. Most of the passengers descended at Sikri, three-quarters of an hour short of Jiri. It was still raining. We continued into the night, curling along the side of the sparsely wooded mountain at barely 15 kmph (9 mph) until there was a loud thud and the bus abruptly came to a halt. We had hit a rock. In the headlamps we could see the beginning of a landslide slowly oozing downhill between the trees onto the narrowest section of the road, where the mountain dropped steeply to the right. The driver reversed. No further, he said. That was it.

Gyalzen jumped out and surveyed the situation. The shoulder was fast crumbling and more debris slithered into the roadway. He came back and ordered the remaining passengers to help heave rocks over the side of the road, thereby clearing a way for the bus. The driver summoned up his courage and moved the Tata forward, using the lowest ratio he could find in his coffee grinder. Once over the rubble we climbed back in, newly drenched, for the final half-hour to Jiri.

The road runs the full length of the village and ends in a clearing on the far side of a cascading stream. Originally we had planned to camp, but there was no longer a question of this. It was past 8 p.m., we'd had no food since 11 a.m. and were tired, cold and wet. More than a dozen lodges lined either side of the road. Gyalzen knew the Sherpa owner of Shakti Lodge, where space was available. We unloaded our gear, moved it under shelter and locked it in for the night.

The lodge owner stoked the cooking fire, adding more wood, and prepared for us large helpings of *dal bhat*, accompanied by countless glasses of *dudh-chiya*, the pre-sweetened milk tea. The rain was still pouring down, but at least we were out of it, in a warm though smokey room lit by a kerosene lamp. We were so wrapped up in our own misery that we didn't see how it could be worse anywhere else. But while it rained in places like Kathmandu and Jiri (1,800/5,905 ft), it was snowing at Pangpoche — in fact everywhere above 3,600 metres (11,800 ft). So while we were relatively snug, trekkers, mountaineers and Sherpa porters caught on the trails of Khumbu, where the storm deposited more than 1.5 metres (5 ft) of snow, faced real hardship and in some cases death that night.

At Island Peak base camp, to the north-east of Pangpoche, an avalanche killed two Britons and their guides, two Tamang tribesmen from the middle mountains west of Kathmandu, as they slept in their tents. Four Spanish climbers disappeared on the south-east ridge of Lhotse Shar, never to be heard from again. Two Sherpas even froze to death on the trails above Namche Bazar, where we would arrive in another ten days. It was a veritable catastrophe. The erosion of the bare mountain slopes caused by the freak storm swept tons of debris into the mountain rivers that would pour, swollen and ugly, into the Ganges, causing silting, flooding and immense property damage. A hundred miles downstream from the catchment areas, Indian peasants would lose their fields and crops as the rivers overflowed their banks, leaving a trail of detritus behind, when anything was left at all.

And while this storm was raging, another storm, nowhere near so deadly, flooded Wall Street with panic sown by greed. We were eleven time zones ahead of New York, so the markets were just opening as we tucked into our *dal bhat.* Over the next few hours, while we slept, the markets would crash, a fall that some said was heard around the world. But the people of Jiri, none of whom had the least notion where Wall Street was, heard nothing, nor did the peasants of the north Indian plain, fighting that night to save their homes and fields, nor those in the wadis of Bangladesh, mourning the countless dead whom the cyclone had washed out to sea. The relentless steamroller of market forces rumbled on, however, and by morning the Dow Jones Industrial Average was 508 points lower, knocking untold billions of dollars off the market values of paper fortunes.

We rose at 6 a.m. to find the storm had passed. The sky was blue, though clouds of steam were creeping across the mountains to the south. Everything was still wet. But the birds were singing and the countryside was smiling when Gyalzen went looking for porters. In addition to the French Baruntse expedition, several trekking groups had waited out the bad weather in Jiri and the demand for porters was great — more than four hundred were needed to clear one hundred trekkers on their way to behold the beautiful Himalayas that, unknowingly and uncaringly, they were helping to destroy.

Each trekker, with porters, guide and kitchen staff, is estimated to consume an average of ten times more wood for cooking, washing and heating than a Nepali traveller, and so the army of more than five hundred people preparing to leave Jiri that Tuesday morning would cut a swath across the countryside on its way to Namche Bazar. During the 10-day trek into Khumbu, the army and its train of attendants would directly or indirectly burn about 15 metric tons of wood. Increase this by a factor of seventy or one hundred to approximate the number of foreigners who visit Sagarmatha National Park each year and the true dimensions of the environmental rape become apparent. It is a catastrophe-in-the-making to which I, too, would contribute and of that I was not proud.

For the moment our attention was focused on coping with problems created by the four-day festival of Tihar, another Hindu celebration that follows the Rama festival of Dasain by two weeks. This year Tihar began on a Tuesday. But as Tihar's first day honours the crow, a bird not especially loved for its role as the messenger of bad news, Tuesday was only a warming-up for the next three days of dancing and feasting. Nevertheless the supply of porters was already scarce as most were heading back to their villages to decorate their homes with flowers and candles.

Tihar is in fact a festival of lights. It is celebrated before the long winter nights descend from the north to darken the plains, bringing winter's chill and slowing the cultivation of crops. Actually its four days of frolicking stretch into five as there is one day of rest before the final celebration, when sisters bestow a red tikha, or dot, on the foreheads of their brothers. While we were confident we could get to the Sherpa village of Jumbesi before Tihar's final day, and there find Buddhist porters to carry us forward, the threat that our progression into Khumbu would be slowed by the religious festivities was a source of concern, as we had the Channel 4 filmstock with us. Thornicroft and Harding were expecting us in Namche Bazar by 24 October, an impossible four day's away. I still hoped to make it within a week, which was going some. It meant, however, that we had to get moving along the trail at any cost, and at any speed, so with the porters Gyalzen was able to hire at a rate of $1.60 per day, by midmorning we set out for Shivalaya, a half-day's walk to the east.

Chapter 3

The Kleenex Trail

Our troubles began at Shivalaya, a peaceful enough hamlet with six or seven houses and tea shops on the left bank of the Khimti Khola, 10 kilometres (6 miles) east of Jiri. No sooner had our porters arrived there early on Tuesday afternoon than they started complaining. The loads were too heavy, they said. Knowing there was a scarcity of porters, they asked more than double the going rate to continue. But what they really wanted was to be back in their villages for the fun-loving four days of Tihar, the Hindu festival of lights, drinking *chang*, playing dice, and chatting up the *didis* (literally 'elder sisters' in Nepali, but in fact a friendly way of addressing any woman).

Gyalzen had heard that sixty porters were coming down the trail from Namche Bazar, so he paid off our complainers and let them return to Jiri, confident he would be able to hire sixteen more without problem. But later he admitted he had underestimated Tihar's importance as a *pujah* — a religious occurrence usually associated with prayer offerings, but for Sherpas, who have adopted this Hindi word, any party or rite, as Sherpas habitually mix feasting and dancing with prayer. While waiting for the new porters, we set up our tents in the clearing by the river and bathed in the sandy shallows downstream. By suppertime none had arrived and, still confident they would appear by morning, we went early to bed. I had a cold coming on, with a raw throat and sore lungs.

Wednesday, 21 October, was 'Day of the Dog', the second of Tihar's holy days. Our objective that day was Kenja, a farming locality of about 120 people on the left bank of the Likhu Khola where it is joined by the Kenja Khola, a fast-running stream that rises near the Lamjura Pass and its neighbouring Phambuk mountain. Kenja is 18 kilometres (11 miles) from Shivalaya. To get there we had to cross the Deurali Pass, descend through Bhandar, a Sherpa village outside Solu-Khumbu, to the Likhu Khola, which marks the eastern boundary of Janakpur province, cross the Likhu by a suspended footbridge and then walk northward along its left bank for 3 kilometres (1.9 miles). Situated at an altitude of 1,600 metres (5248 ft) — 200 metres (656 ft) lower than Shivalaya and the lowest point on our march to Pangpoche — Kenja is the gateway to Solu-Khumbu.

The sixty porters Gyalzen had been told were coming over the Deurali Pass never materialized, and the few that did were intent on gambling their wages away in a local dice game played with shells. Finally we picked up sixteen losers and started over the 2,200-metre (7,220 ft) pass.

The first two things to strike a western traveller along Nepal's trails and pathways is the bustle of inter-village trade, and the absence of power lines, for there is no electricity, which means no television, no electric water heaters and no electric stoves. A transistor can be heard now and then, but not as frequently as in India, for Nepal is measurably poorer than India, the average annual income of hill families being less than $160. Nevertheless the poverty level is not oppressive and rural Nepal seems in general a happier place than the areas I have visited in north-western India.

This is primarily attributable to the fact that although both hill peoples are subsistence farmers, the Nepalese have, for the moment, an abundance of mountain streams from which to draw water and still enough forests for fuel and fodder. They are, therefore, better off in many respects than their Indian counterparts, whose water and forest resources are dwindling. But Nepal's population growth, both from a high birthrate and illegal immigration, is staggering. Before my first trip to the country I had consulted my Rand McNally atlas, 1969 edition, which listed Nepal's population as 9.6

million. In 1986 I was told it had reached 16 million and two years later it was said to be 18 million, having doubled in twenty-seven years. The pressure on the land, 40 per cent of which is mountainous and barely arable, is manifestly enormous. Everywhere I looked new houses and newly cultivated patches were evident.

The fields that flank either side of Deurali ridge were alive with activity that morning as farmers repaired the damage that Monday's storm had caused to the retaining walls of their terraced fields. This was a necessity that had to be completed before they could sew winter wheat. The woods also resounded with the chopping of axemen and we met young girls along the way who were gathering green-leaf fodder for the family cattle, lopping branches off trees to such an extent that eventually they would kill them, while others collected firewood, for everything in Nepali villages is cooked on open fires.

In some places the trail had been washed out, but as we were still in the middle mountains the terrain was not treacherous. At Bhandar, on the far side of the pass, the porters again started complaining that the loads were too heavy and threatened to leave unless we paid them more money. Gyalzen stood firm and grudgingly they consented to continue to Kenja, the last ones trailing into the courtyard of Sonam Lodge at the upper end of the village after nightfall. The lodge is owned by the *pradan pancha*, head of the village *panchayat*, a locally-elected communal council. A tall, imposing Sherpa, he carefully respected the deities and customs of the Hindus who, although comparatively recent settlers on the left bank of the Likhu, were a majority in this part of western Solu.

The usual wage for a porter is 40 rupees ($1.60) a day for loads of up to 30 kilograms. For their 40 rupees, they are expected to feed themselves and buy their own clothes and footwear. Many walk barefoot. They eat a mountain of rice at each meal and wash it down with *dudh-chiya.* With the 40 rupees we paid them they could still get into a dice game at Jiri before the end of Tihar, so they shovelled down their rice and, though it was night, set off at a trot back down the trail to the roadhead. We were left with 550 kilograms of baggage and no porters.

Several things evolved during our unwanted layover in Kenja. The first was that Gyalzen started calling me 'Father'. I never knew whether this was because of my beard, as Sherpas, like Eskimos, have little facial hair and those who do are usually the village patriarchs, or whether it was because of the concern which even before their arrival in Kathmandu I showed for my family of drums.

Gyalzen had begun chiding me about the amount of baggage we had. Inside the barrels were three cameras, an array of lenses, two spotting scopes, one of them electronic, two pairs of binoculars, two Walkman cassette-recorders, a radio, two Thommen altimeters, a compass, some climbing gear, tubes of condensed milk and meat paté, foil-wrapped Fitness chocolate, a wide selection of books, film, batteries, cassettes and toilet paper. But I didn't see how we could accomplish all that we hoped to with any less.

What we hoped to do, of course, was photograph the yeti. Here again I discovered that Gyalzen ardently wanted two things for me. The first was to have me blessed and reblessed by a high lama. I supposed that in his mind this would render me more acceptable to the yeti and to the gods whose privacy and precincts the Sherpas believe it protects. The second was to teach me the Sherpa art of travelling light over difficult terrain, in fact over any terrain at all.

As our travels progressed, that is to say as we had less food to eat, less money for replenishing our stocks, less film to carry, less resistance to the extremes of climate, in short, as our resources ran down, this occurred in any event. For the moment, however, we were entering the third day of Tihar, designated 'Day of the Cow'. That morning the villagers had adorned their cattle with marigolds and buttercups and treated them to extra fodder.

The village of Kenja had not existed twenty years before. It was built by Hindu farmers drawn there by an abundance of wood and water in the lee of the great Lamjura ridge.

Today, Kenja consists of a post office and twenty houses, whose owners live primarily from the trekking trade. Most of the two-storey houses have timbered balconies and overhanging roofs. They are kept neat, some being white-washed, and adorned with geraniums and climbing clematis. As Kenja is on the frontier between Nepal's two major religions, there is an intermarrying of beliefs and practices. While the Sherpas are Buddhists, Tihar is a Hindu festival. Since Gyalzen could find no porters in Kenja he reasoned that in the Sherpa village of Sagar, on the north side of the mountain, he would be able to hire porters willing to work for a decent wage. And so off he went.

Sonam Lodge started filling up in the late afternoon. The newcomers included a British biochemist and her brother, a Yugoslav couple, a young Nepali with his wife and sick child, and an odd trio: a dark-haired girl from New Jersey with her almond-eyed Australian lover and a blonde German trekking mate. The Nepali couple had no money, so the *pradan pancha* let them stay without charge on the balcony overlooking the lodge's courtyard. Their newborn infant was covered in a rash. It cried constantly and needed medical attention, but the nearest dispensary was a two-day walk to the south. I suppose the young husband, who looked not more than twenty, and his wife, about eighteen, hoped to find work in Kenja so that they could earn enough rupees to consult the local *hakim*, or quack doctor.

Gyalzen returned towards nightfall, empty-handed and depressed. He had learned that two large trekking parties — the French from Chamonix and an Exodus group from Britain — were now one day ahead of us and had cornered the porterage market. With our 550 kilograms, we were, quite literally, stuck on the horns of the Lamjura.

That night the *pradan pancha* served everyone chang. His three daughters had decorated the lodge with marigolds, nasturtiums, and rows of candles, and four little Nepali girls came into the courtyard to sing for us. They were shy, forgot their lines, and tried to hide behind each other. Pasang Tshering was fascinated. As the girls were Hindu and he

Buddhist he didn't know the words to their songs but was trying to pick them up. The group led by the New Jersey girl, eating their second dinner on the balcony, were not interested and continued talking through the little concert as if the children were not there.

To attempt to maintain some momentum, I told Gyalzen that I would set off in the morning with the Dutch couple and the three porters we had been able to recruit that day, including the Nepali on the balcony whom we now called Young Husband, and would wait for him in Jumbesi. Although only 16 kilometres away, Jumbesi is separated from Kenja by the Lamjura Pass, which at 3,530 metres (11,581 ft) is the highest point between Jiri and Namche Bazar. It is a long climb over the pass and I wanted to get it behind me. Also, as Jumbesi was a Sherpa village, I assumed we would find all the porters we needed. Gyalzen agreed, but insisted on sending Nawa, the kitchen boy, with us.

As that Friday was a day of rest before the final Tihar celebrations, there was not much traffic over the pass. The climb was longer than I remembered. During the final traverse to the top I started counting the steps. Somewhere between one and two thousand I lost count, and it seemed like another age until I saw the prayer flags in the mist. It had taken me seven hours and twenty minutes, with stops, to climb 2,000 vertical metres (6,562 ft) and I was tired, cold and wanted to reach the tea shop at Tragdobuk to sit by the fire with a glass of *chang* and warm my toes.

Tragdobuk means Under Stone Mountain. It is tucked beneath a sombre rockface and the fields carved out of the surrounding forests are enclosed with split-wood fences. The 22-year-old *didi* who was the lodge keeper looked as handsome as ever. I had lunched there on my 1986 trek to Namche Bazar and she remembered me. After my fourth glass of *chang*, Nawa appeared, so we had two glasses of *dudh-chiya* each before I continued down the valley, leaving him to wait for the others.

Two valleys meet where Jumbesi nestles beside a stream flowing southward from Numbur, a hooded sentinel almost

7,000 metres (22,900 ft) high that guards the gates of Khumbu. The trail from Tragdobuk enters a setting that reminded me of a remote Swiss alpine valley, dotted with herds of small spotted cattle. In the spring the pastures here are covered with primulas, carpeting the countryside in purple, but now they were clothed with pearly everlastings (*Anaphalis*), making it seem like someone had brushed the hillsides with pale green paint. The outlying houses are made of stone with roofs of shingle, while the village houses are stuccoed with mud and manure daubed white, presenting a medieval aspect, like parts of Switzerland might have looked five hundred years ago.

The line of cliffs turn northward where the two valleys join under a rock on which is carved the Buddhist mantra *O Mani Padme Hum.* In moments of stress, Sherpas recite this prayer to the saintly Guru Rimpoche over and over again. They believe it brings them *sonam*, meaning merit, of which one never has enough, and courage, of which they already have a great deal. As I turned the corner into the main valley, the last light was falling and I descended quickly to the Tashi Pal Chenn Lodge. All day as I walked on my own I had noticed the trail was littered with discarded Kleenex, toilet paper, cigarette packs, candy wrappers and plastic separations for pharmaceutical delights. It was not difficult to tell that I was travelling in the footsteps of a large trekking group.

Once in the village I found that not only had the Exodus trekkers taken every camping space, but virtually every spare bed in the lodges, of which there were six or eight. Being less fashionable, Tashi Pal Chenn Lodge still had space available. When Marjolein and Jaap arrived one hour later, they thought it a bit primitive but agreed it would do for one night. Next, Nawa and Young Husband, who was carrying the dreaded Drum No. 8 containing the 35 kilograms of film-stock and some other gear, stumbled in from the dark. Gyalzen had given Young Husband, who was extremely light of frame, the heaviest load. Coming down the stone steps into the village he almost stumbled and would have broken a limb, but managed to catch himself and stood shaking for a few seconds before attempting the rest of the steps. I took his arm and guided him by the light of my headlamp, then eased the drum off his back. His coat and leggings were drenched in sweat. We immediately got him a glass of chang.

Nawa said the other porters carrying drums nine and eleven had come over the pass and would arrive soon, so with my headlamp I climbed back up the trail to wait for them. They also had the Dutch couple's pack and sleeping bags, as Jaap and Marjolein made a point of carrying nothing at all. Although I waited by the big rock for nearly an hour, the two never came.

Even though we were in Sherpa country, the valley was studded with Tihar bonfires, reminding me that tomorrow was the final 'Day of the Didis'. On the way back to the village all the men I encountered were truly well away on chang; they chanted on the terraces in front of their houses and danced arm-in-arm in slow-turning rings, eight or ten of them together, swaying back and forth. The Sherpa owner of Tashi Pal Chenn Lodge ventured that the two missing porters would sleep under a rock that night and wander in early next morning.

Young Husband crossed back over the Lamjura Pass Saturday morning to pick up another load. Since the missing porters had not yet arrived, Nawa and I went looking for them. We met them near Tragdobuk. The old man was limping. He had twisted an ankle and was making only slow progress. We bought them milk tea and told them to rest as there was no sign of Gyalzen.

When finally we returned to Tashi Pal Chenn Lodge, I was depressed, but took some comfort in the fact that Tihar was now over and most porters would be back at work in the morning, their dokos decorated with flowers, all washed and cleaned and in need of cash after almost a week of drinking, gambling and dancing. Jaap and Marjolein also were unhappy. They had discovered that the injured porter and his mate had used their sleeping bags during the night.

Jumbesi is one of the oldest villages in Solu. The Tashi

Thongmon *gompa*[1] in the centre of the village has a sign above its painted door that claims it was established in AD 1636. It is a larger village than Kenja and growing much faster, with new lodges spreading past the seeying-eye *chorten* down to the Jumbesi Khola. North of the village, at a place called Mopung, the Tupden Choeling Monastery was built twenty-one years before by Tibetan refugee monks. As Tupden Choeling contains most of the reliquary and treasure from Rongbuk Monastery on the north side of Everest, with no sign of Gyalzen on my second day at Jumbesi, I decided to fill in time by visiting it. The monks of Rongbuk removed the treasure and smuggled it over Nangpa La, a treacherous high pass on the traditional Sherpa trading route between Tibet and Solu-Khumbu, before their 500-year-old monastery was destroyed by Chinese troops in the mid-1960s. Behind the high lama's chair in Tupden Choeling's main prayer room is a ceremonial headdress crafted in silver and encrusted with sapphires, diamonds, opals and agates. When I entered the prayer room, two monks were reciting from the holy tantras as if talking to a second throne-like chair. This chair, the monk at my side whispered, was reserved for the Dalai Lama should he ever visit the monastery.

The monastery has become an important station along the byways of Tantric Buddhism because it contains a print shop that uses hand-carved wooden blocks to produce many of the most prized Buddhist teaching texts and scriptures which Chinese troops had sought to destroy for all time. The thousands of blocks, blackened with the ink of a dozen lifetimes, are lovingly maintained by the monks, who store them in a room full of shelves on the east side of the monastery's main courtyard. The blocks were their assurance that the corpus of mahayana ritual would continue to survive for at least another dozen lifetimes. But Gutenberg has not yet come to Mopung and probably never will.

Gyalzen arrived that evening in a rain shower. The

[1] *Gompa* is derived from the Tibetan meaning *a place of meditation* and is the word used for a Buddhist temple.

porters, he said, were caught on the Lamjura Pass and would only join us next morning. After six days on the lower reaches of the Kleenex Trail, we still had not got our machine into gear, and I knew the Channel 4 crew would be wondering what had happened to their filmstock. But I was alone in worrying about this. For Gyalzen, our rendezvous with Tihar had been an act of destiny and therefore was out of our hands. As for the Dutch couple, any delay made no difference to them at all. Gaining confidence in their trekking abilities, they were quite happy to dawdle along, taking time to enjoy the magical countryside.

Pasang Tshering accompanied me next morning on the next leg to Manidingma. The day began with bright sunshine and blue skies, and with Jaap and Marjolein we set out for the Apple House in Ringmo, 10 kilometres (6 miles) away. Ringmo is impressively neat. The surrounding countryside is covered with orchards and the villagers make cider, apple fritters and pies to sell to trekkers. Kleenex Trail pioneers had given the biggest house in Ringmo, a three-storey lodging establishment, the name of Apple House, which it now carries proudly on a hand-painted sign over the front door, and it has become a popular stopping place before climbing Tragsindo Pass. We had lunch there and then set out for the cheese factory atop the pass, where we tasted the imitation Gruyère while sheltering from an early afternoon storm. When the rain abated, Pasang Tshering and I ran down the mountain to wait in Manidingma for Jaap and Marjolein. We also expected that Gyalzen and a new complement of porters would arrive before nightfall.

Manidingma has an unsavoury reputation. Many Sherpas avoid stopping there because they believe people in the village practise black magic. Travellers have been known to disappear in Manidingma. Rumour has it that they were poisoned by *shamans* seeking to gain control over errant souls. Once in possession of a dead person's soul, the *shaman* uses it to pursue unwanted rivals or enemies.

On my two trips through Manidingma I found it an uncomfortable place. As much as Jumbesi and Ringmo are tidy, Manidingma appears filthy and rough. The people seem churlish, even though most live off the trekkers. But basically they are wood-cutters and since my last visit eighteen months before they had done an excellent job of sacking the surrounding forests, felling hundreds of pine, rhododendron, magnolia and oak trees. To bark them they roll the trunks onto primitive trestles, and with primitive hand-tools they cut them into beams and planks, exporting the roughly-hewn timber to all parts of Solu-Khumbu.

According to figures cited by the King Mahendra Trust, 400,000 hectares (1,540 square miles) of forest are lost every year in Nepal to timbering, as well as to fuel and fodder gathering. Since the country was opened to trekking in the early 1950s, half the forest cover has disappeared. As the forests diminish, every year the monsoon becomes more devastating, provoking more landslides that clog the river systems with silt. Silting causes downstream flooding and further soil erosion in a constantly expanding cycle of destruction.

Sir Edmund Hillary, who with Tenzing Norkey became the first to conquer Everest in 1953, has warned that 'unless an emergency is declared in Nepal, the hillsides will become deserts within just thirty years'. The figures are terrifying: one in five of the world's population — some 800 million — depend on the sound management of the Himalayas. If the Himalayas are deforested, it is estimated that the resulting environmental changes will turn India into a desert within half a century. This means a human catastrophe of dimensions that would make today's problems of drought and disease in Africa seem minor by comparison. Tourism may be Nepal's biggest foreign exchange earner, but the country's major export is topsoil. Because of deforestation, Nepal loses more than one billion tons of soil each year.

That evening, as dusk climbed from the valleys, there was no sign of Gyalzen. Pasang Tshering began to worry. Calling to Jaap and Marjolein, we moved into Mountain Trekkers Lodge — one of the filthiest places I have ever laid eyes upon. Sipping a glass of *dudh-chiya* near the fire in the first-floor kitchen, I watched a regiment of fleas hop across the vast plain of rumpled blanket on a corner settee. After a *dal bhat* dinner, the four of us were shown to a dirty room and provided with a few dirty blankets. I gave Pasang Tshering my sleeping bag and turned in for an uncomfortable night, feet wrapped inside my empty rucksack and the rest of me draped in a flea-infested blanket that was too small by half.

Gyalzen arrived next morning. Late in leaving Jumbesi, he and the porters had been caught on the pass in a storm. Once again, Jaap's and Marjolein's sleeping bags had provided warm bedding for a pair of porters and the Dutch couple complained loudly. Although showing more confidence, clearly they now regretted having joined our caravan march to Namche Bazar. At this point they refused to go further until assured that their gear would accompany them. But to keep the filmstock moving towards Namche Bazar I could not lessen the pressure and immediately set off for the village of Kharikhola, 12 kilometres (7.5 miles) away.

After crossing the Dudh Kosi by a ropeway and series of tree trunks that replaced a washed-out bridge, most of us reached Kharikhola by mid-afternoon. While Bir Bahadur set up our kitchen beside a tea shop at the far end of the village, I took advantage of the early break to wash clothes in a cold spring by the side of the trail. Returning to the teahouse where we had stored our gear, I found a boy of about sixteen waiting for me. He carried a note from John Thornicroft, the Channel 4 producer, and his programme host, Mike Harding.

Robert,

We *desperately* need film and sound tape for our work. Have been here since morning of 23 October. Can you immediately send us 10 rolls of film, 10 rolls of sound tape and all the batteries.

Also can you double time a porter with the rest of our stuff to Namche. We'll pay the extra. We are absolutely desperate.

Mike Harding,
John Thornicroft.

The boy said he had been trying to find me for two days. He wanted to start back immediately. But this was not possible. Drum No. 8 with the filmstock in it had not yet arrived at the campsite. An hour later, the boy's teen-aged *bahini* (little sister) appeared with a second, more urgent note from the Channel 4 crew, complaining bitterly about the delay and asking to borrow two of our Phoenix tents.

In spite of the distress our delay was causing in Namche Bazar, that night Gyalzen was happy. For the first time since our departure from Kathmandu everything appeared to be running smoothly. Dinner was abundant and good, with fresh vegetables and salad bought in the village. After coffee I went straight to bed, intending to leave for Namche Bazar early in the morning in the company of my little friends, with the film, sound tape, batteries and tents requested by Thornicroft and Harding.

It was going to be a hard two-day walk, following the Dudh Kosi north all the way. The steep mountains on either side of the river had been severely scarred by the rupture in August 1985 of the Langmoche Glacier, to the west of Namche Bazar. A section of the glacier had fallen into a lake above the village of Thame. The lake, so small that it was not shown on the definitive map of the area, Schneider's Khumbu Himal map, overflowed like an over-full bathtub, sending a wall of mud and water into the river below. Because of the debris mixed in with the mud, the wall moved downstream slowly, giving people time to escape uphill, but it swept away dozens of yaks and wiped out every bridge between Thame and the Dudh Kosi's junction with the Sun Kosi, well to the south of us. As it moved downstream, the wall of debris also scraped away forests and hillsides, leaving terrible clefts in the landscape that every monsoon since had widened. The Langmoche rupture provided moving evidence of the Himalayan environment's fragility and showed how the slightest imbalance caused by nature or man was likely to create irreversible damage.

To preserve the forests within Sagamatha National Park, which covers most if not all of Khumbu, the government has forbidden the felling of trees. Exceptions are permitted for the construction of new houses or making limited repairs to existing ones, but in either case a special permit is required from the park warden. Residents of the park are only entitled to gather fallen branches or cut damaged, dying trees. Fines and even imprisonment are the penalty for noncompliance. Visitors, as well as the guides and porters who accompany them, are supposed to use kerosene or butane gas for cooking. The same restrictions apply to the lodge keepers and teahouse operators unless they purchase fuelwood from outside the park. The result is that between Kharikhola and the park entrance at Jorsalle whole forests have been levelled, split into firewood and shipped on the backs of yaks or coolies to Namche Bazar, Kunde, Khumjung and other places along the trekkers' route where it sells for 100 rupees ($4.00) per cord.

Even in the eighteen months I had been away, the progression of tree felling had transformed the timbered slopes of Pharak into fields of stumps. On the hillsides around Kharikhola, the forests are mainly of rhododendron. They are extraordinarily beautiful, especially in spring when the rhododendrons are in blossom. Chipko leader Sunderlal Baguhuna, himself the son of a forester, had told me that it takes two hundred and fifty years for a rhododendron to grow to maturity, which means a tree that stands about 20 metres (66 ft) high, only an hour or two to cut it down and perhaps four days to burn it as fuelwood. Because of the monsoon erosion, these unique but fast disappearing forests would probably never grow again.

In a bend of the trail before reaching Namche Bazar's open-air marketplace there is a teahouse with a view of Everest in the distance. We arrived there around 2 p.m., on Thursday, 29 October, having been on the move since 7:30 a.m. and having climbed during those six and a half hours more than 1,000 vertical metres (3,280 ft). I was knackered. Trudging up the last incline before the teahouse I had my head down, concentrating on every painful step.

'Robert, what took you so long?' I heard someone ask.

I looked up and saw John Thornicroft. He had been waiting there for four days with a retinue of film crew and guides and he was pacing up and down in a state of hypertension.

'We almost decided to cancel the whole thing,' he said 'This delay has cost us £6,000, you know. We'll have to be helicoptered out . . . We won't be able to complete the trek. Where are the ten rolls of film?'

His voice was laden with reproach, as if I had been purposely dragging my feet. I told John that their 35 kilograms of filmstock had been a burden we could have done without. The fact that it had been delayed was certainly regrettable, but the circumstances had quite literally been dictated by the gods.

John didn't understand. He wanted to know why I hadn't left the filmstock behind so that they could have taken it on the aircraft that flew them to the airstrip at Lukla, a day's walk to the south of Namche Bazar along the same trail I had just come over. I reminded him of the instructions he had left for me in Kathmandu.

'We've been waiting six days for you,' he continued. 'It's a total disaster.'

By then I had had enough of his complaining, and so I put on my pack and left without saying a word. Thornicroft caught up with me a while later, still distraught. So much to do and so little chance of salvaging anything, he kept repeating. Then he looked at my rucksack and asked how much weight I was carrying. 'Twenty kilos,' I told him. 'Gee,' he said.

To make sure he understood, I repeated once more that when he had finished with the tents I wanted them returned to Gyalzen Sherpa's home in Pangpoche.

'Ah yes, the caretaker of the gompa,' John said.

'No,' I corrected him. 'One of the villagers who help oversee the gompa's upkeep.'

'All right, I'll have a runner send them to the gompa.'

'Not the gompa, John. Gyalzen Sherpa's home.'

'Oh, right, his house. Well, I have to leave you, Robert, and get back to Tramserku Lodge. We must load up and shoot the arrival of this group of trekkers. Probably I'll bust a gut on the way up to the lodge. Mike Harding is furious.'

And off he went.

I was heading for Pasang Lodge, located at the top of the village, by Mendalphu Hill. Walking past Tramserku Lodge, the teen-aged *bahini*, one of the two messengers sent to find me in Kharikhola, came out onto the balcony and called for me to come in. I thought that perhaps John had remembered he owed for the porterage, not a large sum after all, though every bit helped. But as allegedly I had cost them an extra £6,000, this was unlikely to be the case. The next best prospect I could think of was that perhaps John wanted to offer me a glass of tea and to apologize for his outburst at the teahouse down the hill.

Inside the lodge, Thornicroft was nowhere in sight. Instead there was a short, gnomic person with reddish hair, beard and flashing eyes, who was giving instructions to a Sherpa guide. 'We must film Everest View Hotel tomorrow morning,' he ordered. This was Mike Harding, British TV's good-humour man and a past president of the English scramblers' club.

'Little sister' had obviously made a mistake. She must have thought I was part of the Channel 4 crew and that, unobservantly, I had walked past their headquarters at Tramserku Lodge. Harding, whom I had never met, looked at me as if to say, 'Who the hell are you?' I introduced myself. He flushed and gave me a taste of savage silence. He turned back to the Sherpa. 'I told you to get going!' he barked. When finally his mind had come to grips with my presence, anything really nasty he might have wanted to say was reduced to: 'Why didn't you leave the film in Kathmandu?'

'The instructions were to carry it to Namche Bazar. Believe me, I would have preferred to leave it behind. It created nothing but problems.'

'You've ruined our shoot.'

That I thought was a bit much. 'It seems pretty unprofessional of you to have lost control of your entire filmstock. I assumed you had more with you.'

'We only had two rolls when we arrived here.'

'Well, I did the best I could, in good faith and under difficult conditions. But as it was so important, I don't understand why you didn't fly the filmstock to Lukla.'

'Parajuli said we couldn't.'

I didn't believe that for a moment. 'Then you should not have relied on what someone else told you, but checked it out yourself.'

With that he stomped out of the room. I still had no instructions about what to do with the remainder of the filmstock once Gyalzen arrived with it. And frankly, I no longer cared. I left Tramserku Lodge and, thoroughly disheartened, headed up the hill to Pasang Lodge, opposite the national park headquarters. The climb took another ten minutes, as the lodge overlooked the bowl-like depression that contained Namche's hundred or so ramshackle buildings. Although a grizzly mist had floated in, I had a hot shower — my first since Jumbesi — in the outside shed with a bucket on the roof that served as the shower house and then dressed in clean clothes for a dinner of garlic-fried yak meat. But I was more tired than hungry and, freezing cold, I went early to bed.

Chapter 4

Namche Bazar

The sky cleared during the night and next morning Mendalphu Hill was covered with frost. In the natural amphitheatre below, the houses and hotels of Namche Bazar looked like a toy village.

Namche Bazar's permanent population is maybe five hundred, though it trebles every Saturday for the open-air market. This raucous event, as much social as commercial, is held on a terraced ground at the southern entrance to the village. As the region's only market centre, Namche has become Khumbu's de facto capital and largest metropolis. All paths in the district lead to Namche.

To imagine the setting it is necessary to visualize a bowl sitting on a small table with its back to the wall. The front of the bowl has been broken so that one-third of it is open. Namche Bazar sits in a bowl like this on a high table. Except for the bottom of the bowl Namche has no level ground and the rectangular houses rise up on terraces around its three sides. On the floor at the foot of the table, enclosed in a steep canyon, is the confluence of the Dudh Kosi and Nangpo Tsangpo rivers.

The altitude on the canyon floor is 2,640 metres (8,660 ft). The altitude in the bottom of the bowl is 3,450 metres (11,320 ft). The trail from the Dudh Kosi up to the bowl is almost vertical for the first few hundred metres, with steps carved out of the rockface, then it winds back and forth through a spruce and hardwood forest till it reaches the Everest Tea House, not to be confused with the Everest

View Hotel, which is above Namche on the way to Khumjung. In the bowl itself there are no trees to speak of.

Climbing up the trail on the previous day I had noted that after leaving Pharak the countryside gradually becomes a brownscape. In places there are brown fields and brown pastures. The mountainsides are brown, with brown trees, brown stone houses, and the people are brown. They have little brown children, clothed from head to toe in brown. Only the snowy summits of the tallest peaks are silvery and the sky richly blue. Inside Namche's bowl, brown dust is everywhere: a thin film of it settles on everything and, with each footstep along the pathway, puffs of brown grit rose up my pant leg, soon working into every crease to grate against my sweat-soaked skin.

The trail from the river enters Namche Bazar by the marketplace, at the base of the bowl's south-east shoulder. Past the marketplace the houses are crowded closely together along intersecting paths, with the biggest concentration of them at the back of the bowl. Standing on the terraces of the marketplace, you can see the only common land in the village at the bottom of the bowl: a brown field enclosed by a dry stone wall, with a brown seeing-eye chorten and a heap of brown prayer stones at its lower limit. Looking directly south, out of the broken front of the bowl, the horizon is not distant, for 2 kilometres away the canyon wall on the far side of the Nangpo Tsangpo rises almost vertically for more than 3,000 metres (9,840 ft) to the Kongde Ri ridge.

The backwall behind the village ascends at a 50-degree angle to a shelf at 3,680 metres (12,073 ft). The shelf is called Syampoche and there is an emergency airstrip on it where helicopters and single-engined aircraft can land. Beyond Syampoche the upland continues to Khumjung and Kunde, two villages at the foot of the sacred Khumbilia mountain, but they cannot be seen from inside the bowl. From the marketplace only the jagged summit of Khumbilia shows above the bowl's rim.

On a wooded crest near Khumjung is the Everest View Hotel, an all mod con lodge built with Japanese capital more than ten years before. Because of the lodge's inaccessibility, it has fallen into a rotting, vacant limbo, the wall-to-wall carpeting dank with mould and a cellar that now serves as a garbage tip. According to the hotel's new backers, engineers will soon arrive from Tokyo to rebuild it and the Syampoche airstrip so that, weather permitting, planeloads of tourists will be able to fly there daily from Kathmandu.

Sixty yeas ago, lodge-owner Pasang's mother-in-law had told me, Namche's backwall was thickly forested with tall juniper trees. Today it is bald and rocky, the result of Man's economic necessity and a monument to his unerring stupidity. On the east side of the bowl, the scars of monsoon erosion are vivid. Once broken, the thin crust of topsoil, as it reposes on a sandy base, can almost never be repaired.

But two squares of lighter brown have recently appeared on the backwall, standing out like the designs on a patchwork quilt. They represent Man's belated effort to repair the damage he has done. Namche is well above treeline altitude in the Alps, Rockies or Andes. But here, in the eastern Himalayas, trees can grow, though slowly, up to altitudes of 4,200 metres (13,780 ft). The only problem is that, at this altitude, once a tree is cut down it almost never regrows because of soil erosion and the fact that in so harsh a climate new roots take a decade or more to consolidate.

With funding from Sir Edmund Hillary's Himalayan Trust, the national park staff created several tree nurseries and were experimenting with a reforestation project on Mendalphu Hill, where the park headquarters is located. They enclosed the crown of the hill with a wire fence and planted literally thousands of saplings. Their experiment has confirmed not only that it takes at least ten years for a sapling to set down a root network but that another ten years are needed for the sapling to double in size. As Sagarmatha National Park was established in 1976, the first saplings were only beginning to show signs of robustness.

The experiment, nevertheless, encouraged the people of Namche to undertake a similar enterprise on their backwall. The warden provided them with free saplings and they have enclosed two areas with dry stone walls to keep away browsing yaks, musk deer and Himalayan tahr, a species of wild mountain goat. These areas have become the two lighter backwall patches now seen by travellers as they enter Namche, and though it was too soon to tell whether the saplings had taken root, the villagers hoped that in another sixty years they would again have trees on their backwall.

While some Sherpas have become conscious of the problems caused by deforestation, they are generally not very ecologically minded. Most will argue it is not a high-priority concern, but soon it must become one. In the United States the first communal garbage dump was not established until well into the eighteenth century, long after the Pilgrim Fathers had landed. In Sherpaland, the garbage disposal problem is already acute. When entering Namche from either Thame in the west or Jorsalle to the south, both sides of the trail are littered with refuse, some of it easily burnable like old clothes, mattresses and running shoes, but much of it solid non-biodegradable waste, such as plastic containers and dry-cell batteries.

The Sherpas simply have no notion of waste management. It was not a concept they had to cope with before 1951, when the first expeditions arrived. Until then they had never known non-biodegradable waste. Now they throw their unwanted trash on any hillside, thinking that if it is not in sight of their front yard no problem exists. Burning garbage is out of the question because the foul odours it creates might offend the gods.

After climbing through the litter on the lower approaches to the village, the Kleenex Trail continues past the marketplace to a triangular intersection where it fuses with the old north-south pathway. A few traditional two-storey Sherpa houses remain in the village, generally set back from the main pathways. Their facades are broken by hooded wood-framed windows that once were covered with parchment but are now mostly fitted with glass panes imported from Kathmandu.

The more modern two- and three-storey buildings that front on a main pathway each have one or more ground-floor shops and although not large they stock every sort of merchandise imaginable, from baby powder to rubber boots. But mostly they contain Tibetan curios and the excess goods of disbanded expeditions, such as Bulgarian flashlight batteries, Swiss powdered soups, American freeze-dried meals, British malt tablets, Japanese noodles, cans of Korean tuna, Polish bully beef and even Russian caviar. Climbing gear is another big item. From the latest to the oldest, everything has a price and it's traded in Namche Bazar.

With the trekkers came 'development': a bank, a post office, and electric current from the UNESCO-financed hydro station that works only in the evenings. Add to this a few solar panels, Star beer, Khukri rum, the village's single video machine with a private library of blue movies, yaks instead of street cars, and the only four-storey hotels in Solu-Khumbu — two of them, side by side, at the junction of the main north-south and east-west pathways — and the flavour of 'modern' Namche begins to become apparent. The more popular of the four-storey hotels, New Everest Lodge, is also known as PK Lodge because it belongs to Pasang Kami. Everyone in Khumbu knows Pasang Kami as PK. He is a friend of almost everybody, from every village, but also has admirers from farther afield, including Robert Redford and Jimmy Carter.

Another of Namche's oddities is that, although it sits in a bowl, no stream runs through it. There are open sewage drains in the main pathways, but no water distribution system. The only running water that exists in the bowl pours out of three wooden spouts at the base of a stupa located *at the bottom of the village.*

The best-selling item in Namche is a hot shower. Lodge keepers offer them for 20 rupees (80 cents) a bucket. The water, of course, must be carried from one of the wooden spouts, usually by young Sherpanis (Sherpa women) who hump it uphill in 20-litre plastic containers. Once at the lodge, the water is heated with wood either pilfered from a national park forest or transported from forests to the south of Jorsalle. The only exception is PK Lodge, where shower water is heated by solar panels. As there is no internal plumbing, in most lodges the hot water is gravity-dripped through a sprinkler system contained in an adjacent outhouse. The outhouses are always draughty. If a delay occurs in getting the hot water into the sprinkler system, which sometimes happens, trekkers waiting starkers below have been known to catch pneumonia. Still, there are any number of hardy ones, as I had been the day before, eager to pay good rupees and stand in line for the privilege of risking their health in a Namche douche, a satisfaction that few Sherpas understand, though forever it may bemuse them.

The rest of the water that gushes from the stupa in downtown Namche is channelled through sluices and used to power five prayer mills near the seeing-eye chorten before disappearing downhill into the Dudh Kosi. Why the water doesn't break surface nearer the top of the village I have never understood. But on that last Friday in October I was again pondering this phenomenon as I walked downhill from Pasang Lodge to select a place as near as I could get to one of the spouts to do my washing. The washing area is always crowded; Sherpanis beat their laundry on flat rocks by the side of the stream, rinse and then hang it on the nearest stone wall to dry. It was about 10:30 a.m. and the sun was now warm. I was about to roll up my sleeves and get to work when Pasang Tshering ran up, happy to have found me.

'Father, over there,' he said, pointing some distance above the main pathway to where I could see Gyalzen setting up two tents in the yard of a cousin's house. Just then a tired-looking Marjolein walked by, looking as if she was about to burst into tears. I called to her and asked what was wrong. She came over and said there had been a total breakdown in communications with Gyalzen. This did not surprise me, as lack of a common language aside, they had totally different priorities and I realized now it had been bad judgment on my part to have brought them along. I tried to cheer her up with an invitation to a farewell dinner that night at the home of

Gyalzen's cousin. She looked doubtful, but went to consult Jaap who was waiting at a nearby lodge.

While Gyalzen and Pasang Tshering had arrived more quickly than I expected, it took the rest of the day for the porters to straggle in. Gyalzen became angry when he discovered that in addition to helping themselves to our rice, the porters had pilfered sixteen cans of fruit juice and four bottles of jam. He paid them off accordingly and sent Pasang Tshering ahead to Pangpoche to tell Uncle Pasang he had to hurry on down to meet us with some trustworthy yaks. The plan was to stay overnight in Namche, restock at the Saturday market and leave for Pangpoche around midday.

Jaap and Marjolein checked into the Himalayan Lodge for the remainder of their stay in the Sherpa capital and by dinner that night they were in better spirits. Gyalzen cooked us yak steaks flavoured with garlic and ginger and for dessert we had sliced pineapple soaked in rum.

Uncle Pasang arrived early next morning without the yaks, which were due in later, and after a hurried breakfast the three of us went to the market. Clouds were already rolling in from the east and it was chilly. Gyalzen complained that prices were high because so many trekking groups were in town. The standard Nepali measure is a brass pot known as a *mana*, equivalent to half a litre in volume. The mana has been used in the East since Alexander the Great. As with markets everywhere, early birds command the best prices, Gyalzen paid 15 rupees (60 cents) for one mana of sugar. Later the price more than doubled to 32 rupees ($1.28). Some other prices I noted were 14 rupees (56 cents) for a mana of rice, four rupees (16 cents) for one egg, 7 rupees (28 cents) for one mana of flour, and *suntallahs*, a Terai-grown cross between an orange and tangerine, were one rupee each.

We drank *chang* at the 'Café du Marché', a Sherpa tea house that did a booming business every market day. It was managed by the owner's daughter, whom I estimated was in her early thirties. She looked elegant in a close-fitting aquamarine bodice called a *chunni* and a bluish-purple skirt down to her ankles which moulded her trim form. Her long black hair was held in a chignon by a comb that matched her chunni and she wore black felt Chinese slippers. Her establishment was spotless by Sherpa standards. Porters sat in a line by the windows drinking *chiya* or *chang* and the place was humming with business.

After market we returned to the cousin's home and packed our gear. Gyalzen kept looking toward the rim of the hillside to see if the yaks were coming. Pangpoche was a five-hour walk to the north. Gyalzen decided to wait a while longer with Uncle Pasang and Bir Bahadur, but told Nawa and me to start on our way. It was *bistaari, bistaari* (slowly, slowly) back up to Pasang Lodge, by Mendalphu Hill, where the Kleenex Trail out of Namche resumes. At Sanasa, about an hour north of Namche, the main trail is joined by the path from Khumjung. We stopped there for *dudh-chiya* and examined the Tibetan wares spread on carpets outside the teahouse for trekkers to purchase. As I was haggling with one of the Tibetans over the price of a jade bracelet, John Thornicroft strolled down the hill from the Everest View Hotel. He looked puzzled when he saw me trying on the bracelet and I imagined him wondering why I wasn't accompanied by a slow-moving army of coolies. But if he thought it, he never verbalized it. Instead he announced that he and his crew were having lunch a little further along the trail.

Above Sanasa a yak train was winding its way along the upper trail from Kumjung to Phortse Tenga. It was at once a colourful and a riotous procession, with children laughing and shouting at each other, and the herders encouraging their slow-moving yaks with whistles, torrents of abuse and yelps. Sounds of 'Tchau, tchau!' echoed between the cliffs as they pressed the beasts forward. Some of the yaks had red tassels hanging from their horns and wore bells so that their presence was marked musically as well as physically. With such a crescendo of sound it was easy to understand why no self-respecting yeti was likely to remain in the area when one of these trains passed. Only one that was stone-deaf could fail to hear the approach of so lively a parade.

The autumn countryside was a regatta of browns, dotted by spruce trees with deep blue cones in them, hawthorn bushes turned crimson, and tiny blue gentians bordering the trail. We crossed the Dudh Kosi at the hamlet of Phunki, named after the mountain stream that joins the main river there, and stopped at Nang Lobsang's teahouse for a cup of *chang* before the long climb to Tengboche Monastery. The Channel 4 group was having lunch at the next-door teahouse. Mike Harding came out to have a look around. He didn't say hello. There was not even a hint of recognition.

When I arrived at the chorten on top of the ridge, Tengboche was in the clouds, blotting out the view of Everest, Nuptse and Lhotse, now less than 20 kilometres to the north. The pasture in front of the monastery seemed even more of a madhouse than the Namche market, with the tents of trekking groups pitched in irregular lines across it. Immediately I recognized the two tents we had loaned the Channel 4 crew, with a part of their gear already in place alongside them, guarded by two Sherpanis. I went into Gompa Lodge and ordered *dal* soup with steamed momos, a ravioli-type Sherpa delicacy consisting of tiny balls of vegetable or yak meat wrapped in tsampa pasta, while waiting for Nawa. As he never appeared, after an hour I set out on the last leg to Pangpoche.

The remnants of the 19 October snowstorm became thicker as I dipped behind the north side of Tengboche ridge. The path was slippery and I was worried about getting to Pangpoche before dark, not certain whether I could remember where in the village Gyalzen's house was located. I had stayed too long at Gompa Lodge talking with Nang Wasser, the young monk who supervised its operation. He had told me that the 19 October snowfall had been the worst in fifty years. He confirmed that two Sherpas had lost their lives during the storm trying to get from Pheriche, a village north of Pangpoche, to Phortse, a village directly opposite Tengboche on the right bank of the Imja river — a distance of twelve kilometres — and he also said that the deaths of a sirdar, cook and two expedition members at Island Peak was a tragedy that should never have happened.

The sirdar was Tamang and not Sherpa, Nang Wasser explained. Tamangs come from hill country to the west of Kathmandu and Sherpas consider them less knowledgeable than people from Khumbu about the whims of the high mountain gods. Five years ago the same base camp site had been struck by an avalanche that killed several people. Nang Wasser said the Tamangs should have known this and selected a safer location. He then lifted his eyes toward the hole in the roof over the kitchen hearth, as if silently telling me that the caprices of the snow gods are as whimsical as the smoke curling through the flue.

I met Nawa in the rhododendron and birch forest about half an hour down the trail near the Deboche nunnery. He was coming back to look for me, evidently not having understood when I had told him we would lunch at Gompa Lodge. About eighteen years old, with a gentle, moon-shaped face, he was shy beyond words. The few he did utter were in Sherpa-ka, as he was from Solu, though he knew some Nepali too, but no English. The day was drawing in as we reached Deboche and saw a musk deer standing beside a line of mani-stones. No bigger than a greyhound, it took two bounds, then stopped by a bush and, with ears arched in our direction, looked back at us, deciding whether we were dangerous. It was a gentle encounter. We approach to within six metres before it jumped out of sight, down the ravine towards the river.

Gyalzen caught up with us at the suspended footbridge over the Imja Khola gorge. He was alone, having left Uncle Pasang and Bir Bahadur in Namche as the yaks had arrived late and could only be loaded in the morning. As it was, darkness had overtaken us by the time we reached the village.

Pemba Kunga, Gyalzen's second son, spotted us coming up the trail and with his two sisters came to greet us, jumping with joy. 'Papoo, papoo,' they called. Gyalzen asked Mingma Yanchin, also known as Yangtse, the eldest daughter, where their mother was. Yangtse said that Kanche had gone with the family's two yaks to Lobuche, a day's hike north on the Kleenex Trail, to help her brother, Uncle Nima, bring down a party of trekkers. She had ten-month-old Mingma Ramu with her. Gyalzen hadn't told me they had a new baby.

It felt good to be 'home'. Gyalzen's two-storey house was large and reasonably comfortable. A stable and feed-room were on the ground floor as in old-fashioned peasant houses in Europe. At the back of the dark, low-beamed stable was a steep stairway up to the long room, in which the family lived and cooked. The long room was about 10 metres (33 feet) long, 4 metres (13.2 feet) wide and rose to the rafters, there being no ceiling or insulation, just the main supporting beams and the wood slats to which the roof shingles were attached. In places it was possible to see sky through the chinks in the shingles. Otherwise three roughly carpentered windows along the east wall gave the room light.

The parents slept on a raised bed in an alcove at the north end of the room and near the bed, about a third of the way along the east wall, was a clay hearth set on a layer of slate that served as the family's only cooking range. A banquette ran along the remaining length of the east wall and it was on this banquette that visitors and two of the children usually slept.

At the far end of the room was the family chapel, called a *lhang*, where my bed had already been prepared, also on a banquette, beneath a window that looked onto the peaks of Ama Dablam and Kangtaiga. The windowless back wall was lined with shelves where most of the household material was kept, a good part of it in locked chests. A door by the stairway against the back wall led to a rear porch and the outdoor family toilet that was in plain view of several nearby houses. Like all Sherpa toilets, the excrement fell into an empty space below, and was occasionally covered with twigs and dried leaves, to be removed each spring and used as organic fertilizer in the fields and potato patches, providing precious nourishment for the dry, sandy soil, so fine that the slightest wind kicked up clouds of dust. Animal dung, by contrast, was carefully conserved for fuel.

No garbage was permitted to be thrown into the toilet pit, as the excrement had to be kept ‘clean,’ that is, free of items that might anger the earth and the plant spirits who live under the soil. Such items, as a result, went over the wall at the end of the potato patch or were discarded beside one of the village paths. Nor was the soil fed any compost, as all kitchen leftovers and unused greens were dumped into a slop-filled vat and given to the yaks. Gyalzen called it ‘yak breakfast’ and joked that Kanche sometimes served it to him when she was angry.

As Kanche was not there, Gyalzen cooked supper that evening, first installing me in the place of honour, immediately beside the open hearth. This was both good and bad, for while you receive most of the heat from the fire, you also got most of the smoke before it disappeared through a raised shingle in the roof. It was fiendishly cold and I quickly changed into a polar-fleece suit left behind on my last visit to Pangpoche.

It was Hallowe’en and a half-moon was shining off Ama Dablam, the Matterhorn of the Himalayas, which rises like a hooded priestess with outstretched arms, opposite Pangpoche, on the left bank of the Imja river. After dinner it was immediately to bed under the watchful eye of Lord Buddha and Guru Rimpoche. I thought the sleeping bag would never warm up. But finally it did. In any event, it was a sleepless night as I was still adapting to the altitude. Gyalzen’s house is at 3,920 metres (12,860 ft). I dreamt that a telephone was ringing in the night and the call was for me. It was a curious sensation in this sixteenth-century setting.

Chapter 5

Sanga Dorje's Cave

Next morning Pemba Kunga and little Tsheden jumped onto Papoo's bed and tugged at him until he agreed to get up. Pemba Kunga was eager to show his father the lessons he had learned at school, but Gyalzen's attention was distracted by Tsheden who pouted and screamed until he consented to pick her up and cuddle her. Kunga, disappointed, said nothing, intently watching Tsheden, not quite comprehending why she commanded her father's attention and he could not.

Yangtse, meanwhile, had wrapped herself into a long black *angi*, the ankle-length dress worn by all Sherpanis once they reach the age of 'little womanhood' — not yet adult but old enough to help with all the important household chores. Her first task every morning was to go to the stream at the north end of the village and fetch water. She did it automatically, without being prompted.

I found it odd that, in spite of an abundance of mountain streams, no Khumbu homes have running water. With plastic tubing and the force of gravity, installing an inside water faucet would be the simplest thing in the world, but presumably the arctic temperatures and lack of central heating discouraged people from trying, though I never thought to ask Gyalzen if this was really the reason.

In Gyalzen's home, the situation was even more perverse. Rain leaked through the roughly shingled roof during monsoon, while in winter if ever the roof snow began to melt water unexpectedly started dripping into the chapel or long room. An embarrassed Gyalzen would place a bowl on the floor to catch the water, and shrug his shoulders. Making new wood shingles was time-consuming, and in any event the park warden did not allow the felling of trees for this purpose, so Gyalzen had attempted temporary repairs to the rooftop with plastic sheeting held down by rocks. Aesthetically it was not the most pleasing and, I could attest, not the most efficient either. Like Gyalzen, I too learned to shrug my shoulders. I wondered, however, whether I would survive four long months of this and I was more than a little apprehensive.

I had already noticed that sometimes Yangtse made two or three successive trips to fetch water from the ice-cold stream that cascaded down from the glaciers above the village. As the jerry-can leaked she draped a plastic sheet over her back and carried the 20-litre container with a tumpline. Once back at the house she had to struggle up the narrow stairs with it, and as it was too heavy for her to empty into the copper vessel where the household water was stored someone had to help her. Every Sherpa house has one of these vessels.

Yangtse automatically made the required number of trips to fill the copper vessel to its usual level of around eighty litres. Whenever the water fell below a certain minimum she would set out again to refill it, no matter the weather, rain or snow, or the time, night or day. The jerry-can seemed almost as big as she and just about as heavy, and yet she never complained.

Pemba Kunga's appointed chore, other than attending to the potato patch and gathering firewood on the far side of the Imja Khola, was each morning to bring a bundle of wood up from the stables, while Pasang Tshering lit the fire and helped with the sweeping up. Little Tsheden, as she was only two years old, had no chores. Her only concern was to remain the centre of everyone's attention.

We had a leisurely breakfast of eggs and hash-brown potatoes. Afterwards, while Gyalzen and I discussed plans, Yangtse and Pemba Kunga went outside to the potato field in front of the house and started turning the soil in preparation for the next crop. Pasang Tshering left for Pheriche, a three-hour walk to the north, to meet his mother who was coming down from Lobuche with her brother's trekking party.

Once the sun peeked over the ridge to the north of Kangtaiga the temperature outside warmed up quickly. There was not a cloud in the sky and so I went into the potato patch with a pair of binoculars to survey the terrain on the far side of the Imja Khola. The snow limit was just above the yak pastures of Yaral, directly opposite Pangpoche. I swept the face of Ama Dablam, admiring the 6,856-metre (22,494 ft) pyramid, one of the most distinctively shaped mountains in the world. Its base was less than five kilometers (3 miles) from where I was standing. From Ama Dablam I trained the binoculars on the face of Rawldurche, a 5,200-metre (17,060-ft) mountain above Yaral. Rawldurche was nowhere near so beautiful, forming a long, rocky spine that rose south-eastward towards the Hinku Nup glacier. Suddenly a series of spots jumped out from the snow at me. 'It can't be,' I told myself. The footprints – at least two sets – climbed straight towards Rawldurche's foremost ridge. The snow looked about ankle-deep and I could not imagine what anybody would be doing over there under such conditions.

I called to Gyalzen and he became excited. 'After quick lunch, we go looking,' he said. Himalayan tahr, which Gyalzen called *jharral*, do not remain on that side of the Imja Khola in winter. He explained that the tahr, a shaggy brown mountain goat, moved with the first snow to winter grazing grounds on the right bank of the Imja, where the mountainsides are more exposed to the sun. This convinced him that the Rawldurche tracks had been made by something quite special, like a yeti.

'What about yak?' I asked.

Another glance and he shook his head. 'No yak,' he said. 'All yaks this side of Imja Khola.'

Gyalzen had this curious ability of being able to tell almost without looking where in the vast expanse of brown countryside the yak herds were grazing at just about any time of the day or night. Pangpoche is a village of yak herders and most of the sixty families who live there own pastures in several different places as well as around the village itself, although in and around the village the best land was reserved for potato and buckwheat cultivation. Gyalzen had pastures at Mingbo, Shanjo and Pheriche, all within a day's walk of Pangpoche, and he only had two yaks. His brother-in-law Pasang, married to Gyalzen's younger sister, had pastures at Taboche, Pheriche and Mingbo for his herd of fifteen animals. Also the flanks of the surrounding mountains were used as common grazing ground for the village's combined herd of six hundred yaks.

We set out to inspect the mystery tracks at around 11 a.m. Gyalzen told Ang Dorje, his next door neighbour, and two other villagers we met on the path, what we were up to. They peered across at Rawldurche and hummed seriously, offering a few words of counsel. 'Don't go too high on the shoulder. It might be dangerous,' one woman warned. 'Watch out for falling boulders,' the other advised.

We dropped quickly to the Imja river, crossed it on a wooden bridge wide enough to accommodate one yak at a time, and walked along the left bank to where the Nare Drangka joins it. After crossing the Nare, a stream that rises on Mingbo Glacier, we climbed the escarpment to the pastures of Yaral. Beyond the last walls of Yaral we entered a forest of stunted rhododendron and birch and encountered the first snow. At times it held under us, but more often we sank in up to our knees on the 60-degree slope. With Gyalzen in the lead we fought our way out of the scrub to the first footprints. They were, Gyalzen estimated, about five days old.

One set was clearly made by a primate, for it was wearing boots with Vibram soles. The boot tracks were heading *down* the mountain past us. As we climbed higher, Gyalzen concluded that this set of tracks was made by a 'tourist', because nobody from the village would climb Rawldurche just for the fun of it. About 20 metres (66 ft) to the right, in a snow runnel, were more tracks — quite indistinct but clearly not at all like those left by the Vibram-soled 'primate' nor like the musk deer imprints we had seen in the thicket of trees on our way up. These tracks appeared to have traversed the mountainside from the south, and in the runnel they turned uphill. The snow was reasonably hard here and the imprints were not deep, corresponding to about a size-six foot. The wind and sun had hollowed them out and there were no distinguishable features like toes, hoof or paw marks. From the lee of a large boulder, which obviously had served as a momentary resting place, the footprints led to a bowl about halfway up the mountain. When we reached the bowl, clouds were slipping over the ridge above us, blotting out the sun, and suddenly it became unpleasantly cold.

In the bowl the human footprints converged on the unidentifiable ones and for a while they continued side by side. But when the slope became really steep at the back of the bowl they joined into a single track. We couldn't see how high they went because they became lost among the rocks near the top of the ridge, but we knew of course that the Vibram soles came back down again. The mystery remained complete. We had identified one set of tracks, but not the other. And also we could not imagine what a 'tourist' might be doing on this mountainside. Was someone else searching for the yeti?

At 1:30 p.m., shrouded in cloud, we began our descent. It had seemed pointless to go higher in what had become a total white-out, especially as neither of us was really in shape for a quick climb into the clouds at 5,000 metres (16,400 ft). We returned to the village one hour later. Our cook, Bir Bahadur, had wandered in from Namche and was chopping wood in the potato patch. Not long afterwards, Gyalzen's wife, Kanche, and Pasang Tshering arrived with the ten-month-old Mingma Ramu and two yaks. Kanche unsaddled the yaks, let them find their own way to pasture, then came upstairs and distributed handfuls of candies to the children.

It took a few minutes for us to get to know each other

again. I had bought her three jade bracelets at Sanasa's Tibetan marketplace and some silk scarves from India. She admired them before putting them away in one of the chests and plunging into household work. The snow in some places above Pheriche had been knee deep and she too had caught cold. She brewed a pot of hot raksi for the three of us, mixing it with eggs and rancid yak butter, which is the Sherpa version of an eggnog. It was strangely tasty and warming.

As dusk descended, Kunga came pounding up the stairs and announced excitedly that seven yaks led by Uncle Pasang were coming up the trail from Namche. We went outside to unload them and moved the drums into the long room. Pasang Tshering immediately dove into Drum No. 3, which was a 120-litre job and bigger than him, to distribute the presents from Kathmandu.

After dinner we let the new pressure lamp we had bought in Kathmandu run down and went early to bed, Bir Bahadur and Nawa sleeping on the banquette and the children under covers on the floor. Next morning would be spent unpacking and rearranging the equipment. The children were fascinated by the high-powered spotting scope which we set up in the potato patch. With the scope I was able to follow the set of tracks that Gyalzen and I had not been able to identify on the previous day. They came from a rock band at about 4,400 metres (14,435 ft) on Rawldurche's southern ridge. This had to be investigated, because perhaps they might tell us more about who made them, and so after lunch Uncle Pasang and I crossed the Imja Khola to Yaral once more. Instead of retracing the route taken the day before, we climbed diagonally to the south ridge through the rhododendron and birch scrub, where we came across musk deer latrines. Unlike most animals, the diminutive musk deer, one of the smallest members of the deer family, will walk a kilometre to go to the toilet, always preferring to return to the same spot, as if not wishing to foul other areas of its territory.

After a two-hour climb along the lateral ridge, the afternoon cloud once again started closing around us. By the time we reached the rock band we were in thick soup but managed to find the tracks. Although they were too old to identify positively, they were smaller than yesterday's tracks, suggesting a pair of tahr walking side by side on ground that, though rising, was certainly less abrupt than the terrain we had covered the day before.

This new evidence suggested that our supposed 'yeti' tracks, deep and broad on steep uphill terrain, were in fact made by two tahr, bounding one after the other, placing their feet in the same holes so that they only needed to break one track through the snow. Why these tahr had not followed the rest of the herd across to the Imja Khola's right bank, where the 19 October snow-cover had already melted, was not clear. On this we could only speculate: outcasts, perhaps, an ageing bull no longer able to fend off a challenging buck for leadership of the herd, accompanied by his faithful mating partner? Although we encountered such a pair a few weeks later, we would never know the real answer.

Back at Gyalzen's house I retrieved my short-wave radio from one of the drums and for the next four months it was my only contact with the outside world. Next morning I learned from the BBC World Service that the financial markets had remained in turmoil since our departure from Jiri and that the dollar had plunged to record lows against sterling, the deutschemark and yen. The US Secretary of the Treasury was openly blaming West Germany for maintaining excessively high interest rates and not rehyping the German economy, while the chairman of the Bundesbank said the solution to the turmoil in the currency markets could only be political: the United States must reduce its trade and budgetary deficits and start living within its still considerable means or risk bringing disaster upon the world economy. There seemed to be much tension in the western alliance.

To Gyalzen, of course, the economic squabbling between the great democracies was meaningless. To speak of trillion-dollar budgets and $200-billion deficits was totally incomprehensible to him and, as I was beginning to understand, to a large part of humanity. I paid him 50 rupees ($2) a day for the time we spent together, and kept a 'bank' in the lhang with 20,000 rupees ($800) in it plus some foreign exchange from which he and Kanche could make withdrawls for household needs provided they entered the date and amount in an appropriate pass book. This system worked well because there was nothing in Panpoche — no shops of any sort — on which to spend money. People made their own chang and raksi from the produce of their gardens or the contents of their larders and usually took a jug of it with them whenever they visited friends.

Gyalzen had no idea about foreign exchange or trade imbalances between nations. Provided he didn't travel out of Khumbu, send his children to private schools, repair his house, or buy a new yak, the family could live by barter, which was still as common as currency in Khumbu. As they had quite a bit of grazing land, Gyalzen bartered hay for goods or foodstuffs, depending on what was needed, which never was very much. Sherpas are not big consumers, nor are they intensive users of machinery. Gyalzen had an axe, one mattock, a hand sickle, and that was about it. No lawn mower, power saw, microwave oven or ice-cream mixer. He had a transistor radio made in South Korea, which members of a Korean expedition had given him, a Bulgarian 8 mm. movie camera — which he never used because he couldn't buy film for it and anyway he had no projector, but he proudly showed it to everyone who came to the house — and three large Chinese thermos bottles to keep the morning tea steaming hot.

Gyalzen and Kanche were totally unconcerned and untouched by whether the dollar went up or down, or if the western economies collapsed. They had no television, only some prized religious books from which Gyalzen sometimes recited. As a younger man, he had studied to be a monk. He now got one of the Tibetan books out of the lhang so that he could read to me about the legend of Lama Sanga Dorje. Walking up to Pangpoche from Jiri he had tried to tell me about this wise and virtuous man who is the patron saint of the Sherpas.

Sanga Dorje was the son of Buddha Chendin, a holy man who lived at the foot of the sacred mountain of Khumbilia half a millennium ago, about the time the first Sherpa people arrived in Khumbu from eastern Tibet, migrating over the Nangpa La. Sanga Dorje himself studied for many years at Rongbuk Monastery, on the north side of Everest, but one day he decided to return south to see how the people of Khumbu were getting along. He had earned through his devotion the gift of flight and so he flew over Everest and landed at the village of Phurte, halfway between Namche Bazar and Thame. Unfortunately the people of Khumbu had fallen under the heel of a tyrant, King Tshomtomba, who was camped with his troops at Phurte. The king told Sanga Dorje that his people were not interested in Buddhism and he secretly ordered his soldiers to kill the holy man. Sanga Dorje learned of the king's intentions and fled over the mountains, first to the village of Phortse and then to Tengboche, where he rested on a rock upon which the present monastery is built.

King Tshomtomba commanded his troops to pursue the holy man, so Sanga Dorje fled further up the Imja Khola to a shelter in the highest forest of Khumbu, above the village of Dingpoche. It took the king's troops three years to find the hideout, but when the lama saw them climbing through the trees, he flew to a cave above Pangpoche.

With help from the people of Pangpoche and the neighbouring village of Phortse, a two-hour walk away, Sanga Dorje built the first gompa in Khumbu. The gompa, since rebuilt, treasures among its relics a yeti scalp and skeletal hand. I had already seen these relics, but would go back to examine them more closely later in the winter. The supposed yeti scalp, I came to understand, was more symbolic than genuine, for to the people of Pangpoche it represented a tangible link with their patron saint who had no grave-stone, no final resting place, for when he died his body evaporated into a rainbow. Only his eyes, tongue and heart remained and they were placed in a silver reliquary which now resides at Tengboche monastery.

We decided on our third morning in Pangpoche to visit Sanga Dorje's cave, high above the village, under the austere Taboche Peak, which the villagers believe is the hitching post of the gods. Before leaving, Gyalzen paid off Bir Bahadur and Nawa, who departed immediately for Namche, where the climate was marginally warmer. First we climbed to the Hillary schoolhouse on top of the plateau where the village cremations take place. The schoolhouse was one of a dozen built in Khumbu by Sir Edmund Hillary in the 1960s.

From the schoolhouse we continued up the ridge to the Taboche herders' camp, a cupped depression in the mountainside containing six or seven stone houses and adjoining corrals near the base of the Taboche peak. On the way I thought my lungs would fall out as my sinuses were stuffed and I was still acclimatizing to the altitude. Once we reached the corrals we left the path and began traversing upwards through snow towards the base of Taboche mountain.

At 4,640 meters (15,225 ft) we reached a brownish-grey boulder about half the size of Gyalzen's house under which a dry wall had been built to enclose part of an overhang. This, said Gyalzen, was Sanga Dorje's cave. He went through an open stone porch to the inner part of the shelter and immediately knelt in prayer. He then showed me around.

The dry wall of the porch protected us from the wind, and the view sank down, down, down to the Imja Khola and then up, up, up the slopes of Rawldurche where on my first day in Pangpoche we had seen the mysterious footsteps, still plainly visible, even from these heights. To the north-east was Makalu, Baruntse, and of course Everest, Nuptse and Lhotse. Ama Dablam, the mother's charm box, was almost in our laps, and Kangtaiga was to the south.

At the foot of Ama Damblam were the yak pastures of Mingbo, now covered in snow, and we could distinguish the old Hillary airstrip, used by the World Encyclopedia Scientific Expedition that in 1960 assembled twenty-two scientists and climbers to search for the yeti. Led by Sir Edmund

Hillary, they remained in the field for six months, based in specially constructed huts above Mingbo, and concluded that the yeti was a myth. Behind Mingbo a moranic ridge rose gently to Nare Glacier and the ice-covered wall breached by Mingbo La, a 5,817-meter (19,085 ft) high pass.

I was at first disappointed by the smallness of the cave — two tiny chambers and a semi-enclosed porch — compared to the immensity of the universe outside. For several years, exactly how long Gyalzen was unable to tell me, the hermit monk lived here in meditation. The reality of his world extended no farther than these empty chambers of rock and stone, more than just Spartan, and his mind-expanding devotion to the teachings of his two masters, the Lord Buddha and an eighth-century disciple, the Guru Rimpoche.

Slowly the reality of Sanga Dorje's world dawned upon me. His cave was not restricted to its bare walls but encompassed the entire cosmic universe. He could look towards Pangpoche, and it became another room in his cave, just as Phortse was though it lay 4 kilometers south-west of Pangpoche, on the far side of the mountain. Sanga Dorje's cave *was* the world, the reality of it was as vast as one's imagination. This concept reflected perfectly how Sherpa people think and live their lives. In my own mind I thanked Gyalzen for allowing me this insight.

Beyond the porch, to the right of the entrance, Gyalzen showed me a flat rock that had a hole in it. 'Sanga Dorje's toilet,' he said, almost embarrassed to mention so private a detail of the venerated lama's person. Then we returned to the entrance and sat in the sunshine on Sanga Dorje's doorstep. Gyalzen explained that the lama had been not only a wise and saintly fellow but also a remarkable magician. He used to hang his coat on a sunbeam and kept a yeti for a manservant. The yeti drew water for him and gathered his firewood. Some years after Sanga Dorje moved to this cave the bad king died. Sanga Dorje was able to leave his hermit's abode and he lived for a time in the village of Pangpoche which then had few trees. The yeti used to trim the lama's hair, and wherever the shorn bits fell to the ground a juniper tree grew, which is how Pangpoche came to be the only village in the Imja valley to possess a juniper forest.

About 2 p.m., when the afternoon clouds began to drift up the valley, we left Sanga Dorje's balcony and returned to Pangpoche. Below the Taboche corrals, we watched a family of Himalayan tahr cross the pathway, heading towards a stream. The young goats were baying. Down by the stream, 400 metres (1,315 ft) below, the main herd waited for the late-comers to join them. Gyalzen said that if trapped by a sudden snow storm, whole herds of tahr were sometimes wiped out because, unable to descend to lower pastures, they became snow blind and died.

It was after 3 p.m. when we arrived home. Kanche prepared a churn of salted Tibetan butter tea and as we sat by the fireside sipping it from hooded china bowls, Gyalzen told me more about his family.

Gyalzen was forty-eight, one year younger than I, and in all his life had never seen a yeti, only heard its chattering. He was born in Pangpoche, where his father and mother, the oldest residents of the village, still lived. His father, then seventy-seven, had been a village councillor, yak breeder and artisan engineer, helping to repair local bridges, build shelters and look after the gompa.

Gyalzen had three brothers and three sisters. Only one brother and two sisters survived. As a young man he had been lured by the money to be made carrying heavy loads to high altitudes for international climbing expeditions. He had been several times to the highest camps on Everest and Cho Oyu. He had made two perilous ascents to Everest's Camp VII carrying oxygen cylinders for the Japanese Women's Expedition that conquered the peak in 1976 and had spent a sleepless night alone there, forbidden by the expedition leader to use any of the precious oxygen for himself or to climb higher, so that the summital snows would remain virgin for the Japanese amazons. With the proceeds from high-altitude portering, he had bought the house in Panpoche's lower village.

Kanche hadn't liked Gyalzen being a high-altitude porter.

It was dangerous work and in the last brown pasture before the terminal moraine of Khumbu glacier, with a view of Pheriche valley, was a field of memorials to the Sherpa porters who had lost their lives on climbing expeditions in the surrounding mountains. Kanche did not want to have to erect a memorial there for Gyalzen and so she convinced him to become a trekking guide. Though it didn't pay as well it was safer.

As a boy, Gyalzen had looked after his father's yaks. They had a rock house and corral above the village at Taboche. One December day before Gyalzen became a home owner, his father asked him to take the herd of twenty yaks to Taboche. Here it must be explained that Sherpas believe yetis have supernatural powers. They like to know, of course, that yetis are around and about as they are the servants of the gods, and if the yetis are present that means the gods are at their appointed places. But they also believe that to cross a yeti's path can bring bad luck, serious illness or even death, and to harm a yeti means certain death within two years. Sherpas are, therefore, frightened of the yeti and prefer that the animal stays at a considerable distance. This sometimes poses problems, for the larger variety of yetis likes yak for dinner and has a tendency to run off with any that might stray from the herd, so the yaks have to be protected by means that bring no harm to the rapacious yeti.

Gyalzen told the story of a probable contact he had with several yetis when a young man some thirty years before. Responding to his father's request to take the yaks up to Taboche, with a younger brother and neighbour Ang Dorje, Gyalzen had arrived at the family's Taboche rock house towards nightfall. After corralling the yaks, they shut themselves in and prepared supper before going to bed. They were awakened some hours later by distant noises: Gyalzen described them as a sort of 'Aah-ooh, aah-ooh, aah-ooh.' At first the young Sherpas thought the noise was other herders arriving from Pangpoche with more yaks, or maybe some herders having a raksi party at another rock house because the chattering sounded human. The 'aah-oohs' came closer.

Then someone, or something, tried to force the door as the conversation of 'aah-oohs' continued outside and they heard footsteps circling the rock house.

Until then the boys had not uttered a sound, but when the door started to rattle a second time Ang Dorje fetched some dried juniper boughs stored near the entrance, lit them with embers from the fire, and shoved them through the window. The flaming boughs brought screams from the creatures outside. Yetis, it seems, fear fire. The screams diminished in intensity as the animals fled up the mountain. When the screams were lost in the distance, Gyalzen and Ang Dorje went outside to find all the yaks tightly huddled together, cowering, as they looked towards the ridge from where the noises had last come. Next morning Gyalzen searched for tracks, but there was no snow and he found none.

Chapter 6

Mani Rimdu

Next morning there was great excitement when Gyalzen announced we would take the whole family to the three-day Mani Rimdu festival at Tengboche monastery, where once he had studied to be a *thawa* (monk) and where Pasang Tshering had also briefly gone to school.

We set off after breakfast with the yaks, Marsan and Tsao, who were loaded with two tents and enough food to feed half the village for three days. My cold was slowly descending into my chest and at times I found myself gasping for breath. But the trail was downhill most of the way, save for a fifteen-minute climb from Deboche to the top of Tengboche ridge. The monastery sits atop the ridge at an altitude of 3,867 metres (12,684 ft). We set up our tents in the yard of a monk's quarters behind the monastery, separated from the Chori-Gang meadow by a dry stone wall.

Kanche had cooked special food until late in the night — Sherpa doughnuts, cake and other specialties that would be shared with friends and neighbours attending the festival. We paid a visit to the high lama, Nawang Tenzing Jangpo, who, as he was in the midst of preparations, suggested we would have more time to talk after Mani Rimdu was over.

The high lama looked as if little missed his attention. He was the chief spiritual leader of the Sherpas, who venerated him as the reincarnation of Buddha Chendin, father of Sanga Dorje. Lama Jangpo was born in Namche in 1935. 'At the time,' he wrote in a pamphlet on Sherpa lore, 'there were forests all around Namche and not many people.'

His father, Pasang Nyamgel, was a Namche trader, himself the son of a Tibetan trader who had settled in eastern Nepal around the turn of the century. Shortly after Lama Jangpo was born, his mother took him to visit a great-aunt in Lhasa. They travelled on foot by way of Thame, to the west of Namche Bazar, and up the long Nangpa valley with an unnamed 7,000 metre-high (22,966-ft) table-topped mountain on their left and Cho Oyu on the right. At the head of the valley these two giants are separated by a 1,200 metre-wide dome of ice known as the Nangpa La. This glacier-covered pass, itself 5,716 metres (18,748-ft) high, was used by the Sherpas when they migrated from Kham in eastern Tibet into Khumbu. On either side of the pass the trail is scarred by crevasses and seracs, large castellated blocks of ice that form when the angle of the glacier's bedrock tilts steeply downhill. Leaving Cho Oyu behind them, the family descended Nangpa's more gentle Tibetan side to Tingri, and from there they joined the pilgrims' route to Lhasa. In those days the 700-kilometre (440-mile) journey took a full month, the mother carrying her child in a wicker basket attached to a tumpline, nursing him along the way.

When the great-aunt saw the infant, she was immediately convinced he was a high reincarnate and insisted that the family make a pilgrimage to Rongbuk Monastery. It took them another month to arrive at the world's highest monastery, near the snout of the 15-kilometre glacier that curves down from Everest's north face, and when the local abbot saw the child he needed no persuading. 'This is the true Tengboche *thulku*,' he is reported to have said.

But the reincarnation had to be confirmed by the monks of Tengboche. So the family returned across the Nangpa La to Namche Bazar and when they arrived back home they were visited by a delegation from the monastery. The delegation brought with it some possessions of the previous abbot, who had recently died, mixed with items belonging to other monks, and the ensemble was presented to the infant. Nawang Tenzing was said only to have shown interest in those items belonging to the former abbot, discarding everything else. He was immediately acclaimed the new high lama.

At the age of nine, Nawang Tenzing made a second journey over the Nangpa La to study at Lingbu Monastery in the Tibetan city of Gyangtse. Three years later he transferred to Lhasa's Johkang Monastery and visited the Potala, residence of the Dalai Lama. He only returned to Tengboche when he was a young man and took over management of the monastery from the then-serving regent.

Preparations for Mani Rimdu had been in progress for two weeks. They follow the same pattern every year. The monks spend the first week seated cross-legged in their ochre-coloured gompa consumed in day-long prayer sessions, after which they go to work in the monastery's kitchen for three days making *tormas,* the food offerings to the gods and demons that have set shapes, sizes and colours. After blessing the tormas, the monks continue their preparatory prayer sessions for another four days.

The dance spectacle always begins on the sixteenth day of the ninth month in the Tibetan lunar calendar, after the autumn harvest. People stream in from the surrounding villages to receive the high lama's blessing, an incantation evoking Guru Rimpoche's protection. Gyalzen was eager that I should share in this invocation, as the approving eye of Guru Rimpoche seemed essential to the success of our venture.

Guru Rimpoche was an Indian prophet who introduced Lamaism to Tibet in the eighth century. The Sherpas and Tibetans attribute strong mystical powers to him. His Indian name — Padmasambhava — literally means 'born of a lotus blossom,' and he stands higher in the Buddhist pantheon than Lama Sanga Dorje, being regarded as the second Buddha. His idol is worshipped in every gompa and monastery and the Sherpas believe that mere mention of his name can have a strong beneficial effect.

Tibetan in its origins, Mani Rimdu is a medieval mystery play that celebrates the victory of Padmasambhava, lord of compassion, over the malignant demons and black-hatted fiends who, before Buddhism was established, dominated Tibet. It is a Himalayan variation of the epic battle between good and evil, based on seventeen major confrontations between the dancing monks, each assuming the role of one of the drama's well-known characters.

The dancers enter the monastery courtyard from behind a large curtain covering the main entrance to the gompa. They descend the steps into the roughly paved courtyard and dance clockwise around the *tarchen*, a long cotton prayer flag with sacred tantras printed on it that is attached to a 15 metre-high (50 ft) pole in the centre of the courtyard. Music is provided by two 3 metre-long *dung-chen*, the Tibetan equivalent of Swiss alpenhorns, accompanied by drums, cymbals, conch-shell trumpets and smaller *tseling* trumpets. The music has a measured, rolling, propelling force to it. The dancers live their roles, assuming the spirit and emotions of the fiendish, demonical, ghoulish or godly persons they represent, using expressive hand gestures and controlled body movements.

The first afternoon is a rehearsal for the final day, but without masks or costumes. While the dancers, dressed in their everyday robes of crimson, circled the prayer flag under the watchful eye of Abbot Jangpo, more trekkers arrived in the field in front of the monastery and began setting up their tents. The abbot sat serenely in a special loggia on the north balcony, attended by his marshals and ministers, a silver-hatted *takyo* (tea cup) on a low table in front of him. In spite of the afternoon clouds, a jamboree atmosphere had taken hold and the impression was hard to avoid that the festival, popular as it was among the village people, had become more a festival for trekkers, or 'tourists' as Gyalzen preferred to call them. The tourists were charged a 50-rupee entrance fee and some grumbled bitterly about having to pay this $2 levy. Meanwhile, at the gates of the monastery, Gompa Lodge was doing brisk business. The circus had moved to Tengboche.

The pageant increased in tempo on the second day. More villagers decked in holiday clothes arrived from all over

Khumbu. Our little party was joined by Uncle Pasang, his wife Nima Yanchin, and their three children. We counted more than three hundred trekkers' tents in the field in front of the monastery. After morning prayers the *dung-chen* started making their forlorn, undulating sound and about an hour later a procession led by a marshal carrying a wooden baton and followed by a line of standard bearers exited from the gompa. Behind the standard bearers came the musical instruments: monks beating cymbals; others pounding drums held upright on wooden handles; the conch-shell players; the *tseling* trumpeters; and the pair of *dung-chen* horns, supported in front by a yoke held by a single monk, the *dung-chen* players marching behind blowing their lungs out.

Abbot Jangpo, in a tall red mitre with long lappets, wrapped in gold robes, followed, under a saffron canopy. All other monks wore yellow woollen hats shaped like the helmets of Roman centurions. In front of the abbot was a masked figure with a streaming white beard, high forehead, and bald dome, carrying a long white scarf. He represented Mi-Tshering, a 108-year-old Chinese wise man.

They marched clockwise out of the monastery gate around to the trampled Chori-Gang meadow where an altar had been prepared under a stone shelter. The standards were set in the ground so that they framed Macherma and Taboche peaks. The abbot took his place at the altar and the monks sat in perpendicular rows in front of him. For the next two hours the abbot blessed his monks, distributing 500-rupee ($20) notes to each of them. Then he blessed the spectators. Anyone who wanted a personalized benediction had to make a small cash offering and was given a four-coloured postcard of His Eminence as a souvenir. The children were enthralled by the pageantry. Moreover, they had found many of their friends present and for them it represented a real carnival where fun and religion mingled. The procession reformed as the sun dipped behind the sacred mountain of Khumbilia and they marched back into the monastery.

It was the night of the full moon — so luminous that Everest, Ama Dablam and Kangtaiga shone brightly silver. The

Sherpas danced and sang, drinking chang and raksi, until the moon went down – around 3 a.m. And while they danced, a sheer lack of confidence in the US economy brought the dollar to new lows against other major currencies. In a speech to bankers in London, Nigel Lawson, Britain's Chancellor of the Exchequer, blamed the world market crash on President Reagan's economic policies. President Reagan blamed the Democrats. And the Dow Jones Industrial Average dropped to 1925. In the morning, the real battle between good and evil would begin. Only here at Tengboche the outcome was already known, and for the Sherpas that was reassuring.

Friday's pageant began early. The gallery around the gompa courtyard and the balcony above it were crammed with spectators. The villagers outnumbered the tourists by three to one. That the villagers had installed themselves for the day, not intending to leave until the battle had been won, was clear. They brought with them mats, jugs of chang, thermoses of butter tea and hampers of holiday food which they distributed to everyone around them.

The dancers, with names like Naajam, Tsicham, Tsurcham and the Serpent King, wore brilliantly painted masks and costumes of silk, damask and brocade. They held the Sherpa crowd spellbound; the tourists, however, were restless. Some of the performances required audience participation and often produced ribald laughter. The demons threw bean bags at spectators they didn't like. Ruhrang, one of the worst of the bad spirits from the underworld, teamed up with Hangma, the harvester of death, only to be defeated by Khang Waa, who Gyalzen said was one of the most influential forces of Tibetan Buddhism.

Of the three undisputed stars, one was Dorje Tholla who, although he wore a black mask crowned with skulls and matted yak hair, was described by Gyalzen as a 'good' spirit. He was offered a throne-like chair with a table set in front of it, facing the gompa, and provided with butter tea and tormas. Then appeared Mi-Tshering, the aged wise man who turned out to be incredibly nimble, dragging members of the audience into his antics.

But most popular was Dohlentang, a cave-dwelling hermit. He lectured the Sherpas against abusive drinking, avarice and other ungodly habits, and was attacked by three *tzous*, evil spirits, one of whom resembled a yeti. He defeated them in spite of their treachery, and the audience loved it.

As the sun moved towards Khumbilia, monks passed around baskets of food, including biscuits and fruit, cake and doughnuts. Standing under a balcony was a hazard, as the spectators upstairs sometimes spat, babies were ill or relieved themselves in their mother's laps, and assorted odds and ends were discarded onto those below. But nobody seemed to mind. At mid-afternoon I saw John Thornicroft edging his way towards me. He said he wanted to return the two tents, for they were leaving that afternoon, never having made it to Everest Base Camp. Dohlentang was completing his act by pretending to impale himself on a sword. The audience showed its appreciation by raining coins and banknotes into the courtyard. Monks scampered about collecting the offerings into large wicker baskets. As darkness fell, and the bitter cold returned, the pageant finished with a grande finale.

That night Gyalzen spoke with friends who kept yaks at Chhule, in Nangpa valley, Macherma, in Gokyo valley, and Dingpoche, in Chhukhung valley, and told them he would pay a reward to anyone who reported fresh yeti tracks or otherwise helped us to encounter the animal.

We returned to Pangpoche next morning. The village is built on two levels, like two steps on the side of a mountain. The upper step contains Sanga Dorje's rebuilt gompa, the biggest chorten and the oldest houses, while the lower village where Gyalzen lives has broader, more fertile fields and is more open. The Kleenex Trail runs through the lower village, but to get a better view I had taken the high path after the free-standing gate and climbed towards the gompa. Walking through the upper village, I was stopped by a haggard-looking Austrian climber whose nose was chipped raw by wind-driven snow on the 7,900-metre (26,000-ft) saddle between Everest's south summit and Lhotse.

'Have you seen the Americans?' he asked.

'What Americans?' I replied.

'The Snowbird Expedition,' he said.

Snowbird didn't mean anything to me at the time, though I had seen its graffiti daubed on bridges and stone walls since arriving in Khumbu and wondered what it signified.

'No,' I answered. 'But where have you come from?'

'Everest,' he replied.

'How high?'

'South Col. We were beaten off the mountain by bad weather.'

I forgot about this encounter, until it came back to me three months later. For the moment I was intent on finding my way downhill through the junipers and heaps of mani-stones to Gyalzen's house. It was late and the sun would dip behind the west ridge at 4 p.m.. When I arrived in the long room it was cold, and sitting by the fire waiting for supper of Sherpa stew my fingers were so numb I was unable to write in my journal. Sherpa stew, also called *sharpa*, is made from boiled potatoes, vegetables and blood-red chillies, sometimes with yak meat and tsampa dumplings added. Because it has a warming effect, it is a favourite Sherpa winter dish and for the next few months became one of our staple meals.

Gyalzen had finally caught my chest ailment, only worse. He looked unwell and as if he had fever. I was well aware that if anything serious happened to his health it would likely be blamed on our search for the yeti and this more than anything worried me that evening.

I told Gyalzen his offer of a reward for news of fresh yeti tracks was a good initiative. We discussed plans for setting up a field camp at Donag Tsho, one of two lakes north of Gokyo, as I wanted to test his enthusiasm. It remained unchanged. We estimated that in good weather it would take two and a half days to get there. My friend and co-researcher Gerard Metral was due to arrive at the Lukla airstrip on Monday. It was Gerard's first time away from the Alps and I was concerned that he would be overwhelmed by culture shock, so I wanted to be there to meet him. I decided to take Uncle Pasang with me, leaving Gyalzen to rest and hopefully to recover. Other than wide-spectrum antibiotics, our medicine chest contained nothing that would reach down into his chest and I was scared that strong antibiotics might upset his system, doing more harm than good.

I left with Uncle Pasang on Sunday morning, 8 November, for the two-day hike to Lukla. Pasang was no slow-poke and when I arrived at Everest Lodge in Lukla on Monday evening I was exhausted. Gerard was waiting, but I could tell by his mood that something was wrong. He seemed blue and, speaking no English, felt cut off from the rest of the world. Once he saw me, however, he never stopped talking. Gerard had with him one bottle of Ricard, the licorice-tasting *anisette* alcohol, another of Génépi, a liquor distilled from alpine herbs, and with the lodge keeper's supply of chang and hot raski there was no chance of us going dry that evening.

Gerard was a bearded, thirty-eight-year-old postman. His job was to deliver telegrams, so he ran all day long up and down the Chamonix valley in a little Renault pick-up truck painted in the yellow and blue livery of the French postal administration. The job was beginning to bore him, for he had an inquiring mind, being a tinkerer, and a roving eye. His problem, I quickly learned, was that he had fallen newly in love, but not with Josiane, his live-in girlfriend of more than a dozen years. He had left his heart with Laurence, a 24-year-old postal employee who had recently moved to Chamonix from Paris. Their affair was consummated two nights before he left for Kathmandu, and although he had taken a three-month leave of absence, he was already dreaming of returning to Laurence's side. I doubted he would last the winter. He didn't like the raksi that much. Jokingly he said that when the bottle of Ricard was finished he would leave. As I was not an anisette drinker, but quite enjoyed the raksi, I thought the litre-bottle of Ricard would last at least until the New Year.

Chapter 7

Phortse

We left Lukla at mid-morning for Namche, having hired a 14-year-old Sherpani to help carry Gerard's gear. I was not feeling well: my legs were weak, a warning that I was going to have a difficult time on what normally was an easy trail. Gerard, in good form, strode ahead and there was no possibility of my keeping up with him, even though theoretically I was acclimatized and he not at all.

The day was drawing in when I reached the bridge over the Dudh Kosi, where Gerard was waiting, to begin the steep climb to Namche. His T-shirt was dry, mine clinging to me like a wet Kleenex, and I had already changed twice. Above the gorge I had to stop every few steps. It felt like a weight was boring into the right side of my chest. When I made it to the hairpins, I had no strength left in me. I was mildly concerned about passing out and falling over the cliff, but otherwise the sensation was dreamlike, sort of floating outside of and beside my very tried body.

Pasang came down in the dark to rescue me. He took my rucksack and guided me up to the Everest View teahouse. There was no question of my going further. I was cold, dripping wet and weak. We made arrangements with the owner to let me sleep on one of the benches for the night. He wanted to serve me food but I knew nothing would stay down and requested instead a thermos of lemon tea, for I was completely dehydrated.

Gerard and Pasang continued to Pasang Lodge beside Mandalphu Hill, to come back and fetch me in the morning.

I barely had enough strength to change into my polar fleece and climb into the sleeping bag. But once inside the down bag I slept for twelve solid hours.

Next morning I was feeling much better, though still weak. After a breakfast of tsampa, I walked slowly up to Namche, but decided I would not be going much further that day. A lead weight was still in my chest, which was painful. So we spent the day resting in Namche, arriving in Pangpoche only late in the afternoon of 12 November.

Kanche was pleased to see us, but Gyalzen looked ill and seemed worried. It was the last night of the autumn *Nyugne* ritual and Gyalzen said eighteen people were fasting in the village gompa. *Nyugne* is a soul-cleansing rite held twice a year for those wishing to scrub away some of the sin accumulated between seasons and earn new *sonam.* Gyalzen lit three candles in the lhang and said prayers. I think he would have liked to have been at the gompa with the others seeking atonement, for the sound of the drums and cymbals floating down the hill seemed to distract him.

Gerard and I spent the next few days preparing our equipment for the first sortie to Donag Tsho and chasing tahr on the mountainside above Pangpoche where we counted two herds of about forty animals each. They resembled a cross between a large chamois and a bouquetin. Whenever we got to within twenty metres of them they sought the protection of the nearest ledges where they knew we could not follow.

On Monday, 16 November, Gyalzen walked with us to the summer pastures of Mingbo. This daylong trek was intended to determine whether I had recovered sufficiently to set out for Donag. After crossing the Imja to the plateau of Chhulungche, we climbed the old moraine on the right bank of the Nare Drangka. It was well exposed to the sun so there was no snow, but on the opposite bank winter was firmly installed. We noticed that our progress was being observed from the crest of the moraine by a large male tahr with great curved horns. He was accompanied by a cow. As we approached to within one hundred metres they sauntered down to the river, forded it and climbed the far side till they reached the snows of Rawldurche.

It was fascinating to watch them work, the male leading on an almost vertical wall of moranic gravel scraped bare by a glacier rupture ten years before. He was a handsome creature, with a long, shaggy coat, and obviously quite elderly. Gerard estimated he weighed more than 150 kilograms. When they got into the snow well above the river they encountered a new set of problems. The snow was deep and the stag had to plough a path through it so that the cow could follow. He would rise on his hind legs and lunge forward, plant his forelegs, then follow with his hind legs into the same hole in the snow, and off he would bound again, leaving only one imprint for all four legs. It was easy to see under these conditions — uphill, in deep snow — how the tracks of a four-legged tahr could be mistaken for a bipedal whatzit. The female followed in exactly the same tracks, enlarging them slightly, with the result that it seemed like a quite considerable two-legged creature had passed that way.

'Very good climbers,' Gyalzen commented. And indeed they were. Also they had solved for me the mystery of the Rawldurche tracks and for that I was grateful.

Our interest in Mingbo was due to several factors. After a late snowstorm in April 1986 Uncle Pasang had gone there to gather fodder for his yaks. Arriving at the head of the valley he saw what he thought were fresh yeti tracks. The animal had slithered down the mountainside and then the tracks disappeared behind some nearby rocks. As the tracks seemed not more than a few hours old, Pasang supposed the animal was somewhere close by. He fled back to Pangpoche and warned other villagers not to go to Mingbo as the yeti was roaming about.

Mingbo had been a centre of yeti activity for many years. But aside from Uncle Pasang's experience — which I surmised may have been the tracks of a tahr or some other non-yeti creature — there had been, as far as I could discover, no recent yeti encounters at Mingbo, the last ones having occurred almost forty years before.

In March 1949, one of Gyalzen's neighbours by the name of Mingma had taken his yaks to Mingbo and late one afternoon heard what he thought was someone calling him from the ridge above. He answered and was shaken when a yeti appeared among the rocks. He described the animal as being about the size of a teenage boy. Mingma hid in his stone house but the yeti followed him there and waited outside. Mingma was terrified. He lit a fire in the chimney and heated an iron till it was red hot. He poked the glowing iron out of the window and to his relief the yeti ran off.

During the winter of 1950, another herdsman from Pangpoche by the name of Dakhu was tending his yaks in the pastures behind Mingbo. One afternoon he noticed that a yak was missing and he went in search of it. About thirty metres away, half-hidden by the rocks, he saw something hairy which he mistook for his yak and called it by name. The animal suddenly stood up and came slowly toward him on two legs, pulling up tufts of grass and lobbing them at him. Dukhu fled back to Pangpoche and told his neighbours what had happened. He described the yeti as being small like a boy with a pointed head and a body covered with long reddish-brown hair.

Both incidents involved a *mitey*, the smaller of the two types of yetis believed by the sherpas to inhabit Khumbu. The mitey — *mi* being derived from the Tibetan word for man and *tey* being a generic word for animal, hence a man-like animal — does not attack yaks, though some Sherpas say it is aggressive with humans and will kill anyone who comes too near. It supposedly survives on a diet of rodents, such as marmots, mouse hares, voles and pikas, mixed with wild tubers, roots, berries and, if it can find any, frogs.

Other Sherpas have told me a mitey cannot actually be seen because it is god-like. I'm not sure what precisely they meant by this. Perhaps the explanation of Gyalzen's Bodnath brother-in-law, Lobsang Tshering, was closer to what they felt. Lobsang once told me, 'A mitey sees everything, but can only be seen if it wants to.' Their natural camouflage is so perfectly suited to the reddish-brown rockscapes above

4,000 metres, particularly in the scree and rocky beds of former glaciers, that they would be almost impossible to see anyway, even when they moved, unless silhouetted by snow or sky, or unless they wanted to show themselves, which may have been the sense of Lobsang's comment. In any event, Gyalzen thought Mingbo might make a good alternative site for a yeti stake-out if ever Donag Tsho proved unsatisfactory.

The plan we had agreed upon remained unchanged, namely to establish a field camp at Donag Tsho, a glacial lake surrounded by towering peaks that looked like church spires, about three-quarters of the way up the Dudh Kosi valley, immediately to the west of Pangpoche. From there we would prowl the surrounding countryside throughout the winter months. From the field camp we intended to set up a regular circuit that at least once a week we would travel to observe what was going on in our 'territory'. I was increasingly concerned, however, that Donag Tsho would be problematic because I had learned in the meantime that the upper Dudh Kosi valley was now kept open for trekkers during the winter months — at least as far as Gokyo — though, of course, there was less traffic in wintertime than during the high trekking seasons of spring and autumn. In case anything went wrong at Donag I wanted an alternate site for a field camp and thought that Mingbo, being easy to supply because of its proximity to Pangpoche — that is, a two-hour hike away — could possibly fit the bill.

Mingbo is a curious place, three rock houses and a few rock camps located on a narrow shelf at an altitude of 4,520 metres (14,830 ft), between the Nare Drangka and a rock-strewn ridge near the base of Ama Dablam. Tracks of animals, big and small, were everywhere in the snow when we arrived there, but they were so old and distorted that it was impossible to tell whether man, or which beast, had made them. Following one set of tracks, I wandered behind the large rounded hill to the east of Mingbo, heading in the direction of the 1960 Hillary airstrip, pleased that I seemed to have recovered my form and was again acclimatized to the altitude. I had found that after an illness, especially if antibiotics had been part of the cure, one had to rebuild a tolerançe to altitudes above 4,000 metres (13,125 ft).

The tracks had become obliterated in the rocks when Gyalzen caught up with me. I was startled to note that his face was drawn and it seemed as if he was not getting enough air into his lungs. His voice was small and, on the way back down, his footsteps were unsure. He lagged behind and was in bad shape, similar in every respect to my washout of a few days earlier. So it was clear that we could not leave the next day, as we had hoped, for Donag Tsho.

Gyalzen was depressed by this new delay. If something was not done to improve his health and spirits I feared he would think the curse of the yeti was upon him. As the Himalayan Rescue Association maintained a clinic at Pheriche, a half-day's walk to the north, that was staffed every trekking season by two doctors and a nurse, I decided to coax Gyalzen to visit it. But he was reluctant. I pointed out that we had to go to Pheriche anyway to pick up four insulator mats needed for our Donag Tsho excursion. Kanche owned a small house there, transformed into a lodge, which she opened only during the high season, and it had a supply of insulator mats. This he accepted, recasting the outing to Pheriche as a family picnic. With Uncle Pasang and Kanche, who carried little Mingma Ramu in a basket on her back, we set out next day. The weather, as long as the sun was not shielded by one of the valley's many peaks, was summerlike.

Six kilometres north of Pangpoche, Pheriche sits astride the Kleenex Trail at the southern end of an alluvial upland watered by the Lobuche Chubung, a small glacial stream that empties into the Imja Khola. The gently rising upland that stretches north of Pheriche to a millenia-old mound of glacial rubble at Duglha, 3.5 kilometres away, is lined on the east by the remnants of a well-rounded moraine and on the west by the granite pillars of Taboche and Tshola peaks. In the northerly distance we could see Cho Oyu, the world's seventh highest mountain, as well as Lobuche peak and Pumori. Everest was hidden from view by Pokalde. Pheriche, at an altitude of 4,243 metres (13,921 ft), rivals Dingpoche, which is 40 metres (131 ft) higher but only sporadically inhabited during winter, as the highest permanently settled village in Nepal.[1] It counts perhaps twenty houses and lodges, spread out because of the opening onto the marshy upland which in summertime becomes a vast pasture.

Kanche's lodge was a modest single-storey structure, with an open-hearth kitchen to the right of the main entrance, and a dormitory to the left. Its front yard was separated from the pathway through the village by a stone and peat wall. A peat-covered bench ran the length of the yard inside the wall.

The Himalayan Rescue Association clinic counsels trekkers on how to adapt to high altitude. Severe altitude sickness, if not quickly diagnosed, can be fatal. As the clinic is located only 50 metres from Kanche's lodge, before lunch I wandered over and met Dr. Wade Henrichs, a thirty-year-old specialist in emergency medicine from Minneapolis, Minnesota, his wife Paulette Bergh, and nurse Mary Anne Hauser, from Durango, Colorado.

Four thousand climbers and trekkers visit the clinic each year. One in three persons who journeys up the Kleenex Trail is affected by problems related to altitude sickness. It is something that can strike anyone — Sherpa or tourist, climber or trekker. The most common symptoms are headache, nausea, loss of appetite and retention of body liquids. Its worst forms are brain and lung oedemas. Oxygen and some drugs like Diamox help, but the only sure cure is to lose altitude as quickly as possible. The clinic has been credited with saving dozens of lives. In addition, it provides virtually free medical services to the local villagers.

An American climbing team that had been defeated on Ama Dablam was gathered outside drinking beer and testing Mary Anne Hauser's medical talents. As I explained our medical problem to Wade, Gyalzen, curious to find out what

[1]The highest permanently inhabited village in the world is in Bolivia, at 5,335 metres (17,500 ft).

was going on, wandered over with Mingma Ramu tied in a blanket on his back. I introduced him to Wade and Paulette. Wade prescribed a five-day mild antibiotic treatment for Gyalzen, producing the medicine from the dispensary, for which he charged a nominal 100 rupees ($4). Moreover, Gyalzen was told it would be wise to start taking the antibiotics immediately. Gyalzen, who had been out of breath on the trail that morning, meekly complied.

To be on the safe side, Gylazen asked the Pangpoche lama to come by the house that evening and say some special prayers. The lama arrived after dark with his dog and while he was fed dinner consulted with Gyalzen on which tantras should be used for the case at hand. Finally they selected a book from the lhang and the lama began to chant from it, sprinkling rice about the room and burning juniper and incense which he required Gyalzen to inhale. He then made a forest of small tormas with tsampa dough, including an effigy of Gyalzen. He placed the figures on a tray which after his recital of the holy texts he took downstairs and placed near the entrance of the house to encourage the evil spirits to leave.

Depending on Gyalzen's health, we planned to leave in the morning on the first leg to Donag Tsho. The trail to Donag followed the right bank of the Imja southward to Phortse, then headed north above the Dudh Kosi's left bank to the terminal moraine of the mighty Ngozumba Glacier at Nah, crossing the river beneath the moraine, climbing first to Gokyo and finally to Donag. The total distance of 25.5 kilometres (16 miles) required two and a half days of travel, with a gain in altitude of 1,000 metres (3,285 ft).

In addition to two yaks, we decided to hire three young Sherpanis to carry food and gear for a two-week sortie. Still working on the principle that we would make our main operations base at Donag, I proposed leaving a stock of material there at the end of our sortie while we returned to Pangpoche for more supplies. Tentatively I had suggested that we use a summer *yersin*, or yak herders' rock camp, in the lateral moraines to the east of the lake as our permanent camp. Both Gyalzen and Uncle Pasang thought this a good idea.

As we sat by the hearth that evening, Gyalzen went through his repertory of yeti stories. Tenzing Norkey's father, originally from Tingri in Tibet, settled in Thame many years ago and became one of Khumbu's largest yak breeders. He had pastures everywhere, Gyalzen said, but always kept a large part of his herd in Tshola Khola, a lateral valley between the Dudh Kosi and Imja, on the east side of Tshola pass. One monsoon season a yeti known to roam that region made its appearance and started stalking the yaks. When the herdsmen attempted to frighten it away, the yeti perched on top of the rock under which they were camped. The herdsmen became afraid and built a semi-circle of fire at the entrance to the rock overhang to prevent the yeti from attacking them during the night. They threw all the fuel they had onto the fire and the flames leapt so high that the yeti, chattering madly, fled into the mountains and never bothered them again.

Gyalzen also told us about Ang Dahki, a Sherpani whose parents live in Pangpoche but own a stone house at Dingpoche and pastures beyond, near Chhukhung. Late one afternoon in April 1986 Ang Dahki went to look for the family yaks as they had not returned home. Where the Chhukhung valley narrows between a moraine and the mountain she heard what she thought were the screams of several yetis. When she stopped, wondering what to do, three large rocks cascaded down the mountain from where the screams had come. The rocks narrowly missed killing her. Terrified, she ran back home and sat mute by the fire for several hours before her parents could coax from her what had happened.

Gyalzen was a good story teller. He enjoyed the sense of drama he could create with his tales. And he could also be quite amusing. Story-telling is still considered an art in Sherpa society as there is no television and little radio. The Sherpas can spin yarns for hours on end, seated around their hearths. For the most part, the listeners remain wrapped in the anecdotal dramas till it comes their time to add a detail or to branch into another tale. Characteristic of these sessions was the ease of mirth, for Sherpas take great pleasure in laughing, not only at others but also at themselves. I regretted on such occasions that I spoke so little Sherpa-ka.

Gyalzen mentioned that in addition to screaming and chattering, yetis sometimes whistled like humans. He also remarked that the yeti, when it travels, usually travels at night, which was something I had already concluded from my research. But he added rather pointedly that the yeti when it travelled was not encumbered with baggage, giving it a distinct advantage over us. He was eying the gear we had assembled for next morning's departure to Phortse.

I was now concerned by Gerard's declining morale. Whereas at the beginning he had been impressed by Gyalzen's home, he now found nothing pleasant to say about it. He complained about the dirt and the seeming unwillingness of the people in Khumbu to improve their living standards. He considered the Sherpas to be hostages of a regressive religion and he failed to see why they couldn't understand this.

Next morning as we completed our packing, Gerard broke into a tantrum when he discovered we had no more black plastic bags. He had intended to use them as solar-heaters for water at the operations camp. We had six of them at the outset but I explained that the porters had ripped them apart for tenting when they were caught in storms on the Lamjura and Tragsindo passes during the caravan march to Namche.

'Nothing is as it's supposed to be,' he suddenly lamented. 'Everything has fallen apart. First we were supposed to be an expedition. Now we're just two, on a trek. We haven't got all the necessary equipment and we can't even afford enough porters. We have no fixed base camp as you said we would, and no rented hut at Gokyo. Your organization just isn't serious.'

Some of what he said was true. Initially I had hoped to rent winter quarters at Gokyo, but these plans were abandoned for lack of funds and also because we had since confirmed that Gokyo was no longer deserted in winter, as once it had been. Several lodges remained open to accommodate the Japanese and Australians who for some unknown reason prefer to do their trekking during the winter months. But still, because of the snow I hoped all trekking traffic would stop at Gokyo, normally the end of the line, and we would be left in peace at Donag Tsho, one lake further north in the Dudh Pokhri chain of five glacial lakes. *Dudh* in Nepali means *milk* and *Pokhri* are *lakes.* The Dudh Kosi, which exits from the Dudh Pokhri, is therefore the milky river.

The trail to Phortse begins by the chorten in the upper village and follows the right bank of the Imja, above and opposite the Tengboche ridge. Phortse is claimed to be the first village in Khumbu to have grown potatoes. There are various stories as to how and when the English sahib's tubers were introduced to Sherpa culture and came to form the staple of their diet, but it seems that in the nineteenth century an enterprising young Sherpa engaged as a gardener by the resident British minister in either Darjeeling or Kathmandu pocketed a few and took them home with him on his next visit. It was like magic. Potatoes adapted well to the altitude and Khumbu's sandy soil, and Sherpas adored their taste, a welcome departure from the ground buckwheat tsampa, which sometimes they mix with corn or barley flour.

Phortse has no lodges or teahouses, although some homes take in trekkers for a fee. The village is situated on a triangular plateau above the confluence of the Imja and Dudh Kosi rivers, at an altitude of 3,840 metres (12,615 ft). Behind it a backwall protects the village from northerly winds, giving it a micro-climate. Its thirty or so houses are large and well spaced apart. They all have first-floor terraces off the long room. The terraces are half-covered, face south, and are used for drying grain and corn.

Phortse people are yak herders. A few goats and sheep, relatively rare in Khumbu, also amble about the village, and dogs as well, lots of dogs. But no chickens. In fact I have only once seen a chicken in Khumbu, at Namche, and I asked the reason why. Sherpas love eggs, and great numbers are sold at the weekly market. But Sherpas consider the parents of those eggs to be unholy — the rooster represents vanity and is depicted as one of three corrupting forces that grease the wheel of life. No chicken is therefore permitted within sight of Khumbilia, residence of the god Khumbilia Terzen Gelbu, who I am told is one of the twenty-one foremost gods of Tibet. And so the eggs consumed in Khumbu are carried there in the sacks of the grain, rice, flour and sugar by the merchants who attend the Saturday market. In any event, the biology of chickens is such that they are not efficient layers at altitudes above 3,000 metres.

Tantric Buddhists believe that Guru Rimpoche, the Indian mystic who reformed Tibetan Buddhism after travelling to Lhasa in 747 AD, subdued the original Bön gods who ruled the high Himals and converted them to the cause of Buddha. His most powerful weapon was the *vajra* (*dorje* in Tibetan), symbolic of Indra's thunderbolt. Guru Rimpoche also visited Khumbu, and with his magic vajra and an ample bag of tricks vanquished the Bön gods and devils he found there, sparing only those who consented to guard and protect Khumbu's valleys as a Buddhist sanctuary.

Sherpas, who migrated from eastern Tibet about the time of the Mongol invasions — hence their name of 'People from the East', *Shar* meaning east and *pa* meaning people — believe that every major mountain is the home of a god. Khumbilia, for whom they named their country, was by far the most revered and kindly towards them, and this ever since the days when Buddha Chendin, father of Sanga Dorje, used to entertain Khumbilia at his hermitage. In Sherpa mythology, Khumbilia rides upon a red horse and looks powerful, with a pale face and white robes. Khumbilia's wife, Tamosermu, lives on Tramserku, to the south of Kangtaiga, itself named after one of Khumbilia's students who lives there.

Among the most feared of the local deities is Chomolungma, who lives atop Everest. She is the mother goddess of earth and does not like to be disturbed. She rides a red tiger and, with her orange skin, is very pretty to look at. She is surrounded by flowers and dressed in a robe of many colours. She holds a shrew-mouse that vomits wealth in her right hand and a bowl of fruit in her left.

The people of Phortse are conservative and hard working, but they are not wealthy. This may be because they once had an argument with Sanga Dorje, who sought refuge with them when fleeing King Tshomtomba and they forced him to fly across the Imja Khola to Tengboche. Apparently they regretted this and later made frequent pilgrimages to Pangpoche to seek his forgiveness, eventually helping build the gompa there which to this day they share with the villagers of Pangpoche. Phortse has no gompa of its own.

While the people of Phortse may not be wealthy in a material sense, they own many yaks and they are rich beyond compare in three things: mani stones, many of them exquisitely carved, impeyan pheasants, the brightly-plumed national bird of Nepal, and musk deer. There are so many musk deer around Phortse that animal biologist Bijaya Kattel based his field study project there. Since Gerard and I arrived in the village before Gyalzen, Kanche and Uncle Pasang, and therefore had time to spare, I decided to see if Bijaya had returned from Kathmandu. I knew he rented a house somewhere in the village and asked the few people I met if they knew where it was.

I soon found a boy in ragged brown clothes who agreed to guide us to Bijaya's house — near the bottom of the village — where we met Beg Bahadur Baniya, a national parks worker assigned to the musk deer project. Beg Bahadur was about to leave for Saturday market in Namche. He nevertheless invited us in and offered us tea. Beg Bahadur, in his early thirties, was married and had three children but they lived south of Solu, where the climate is much warmer. He said he saw his family once every six months. Yes, he agreed with Gerard, that was a long time, and he missed them. He said Bijaya Sahib was expected from Kathmandu in two weeks. As Beg Bahadur was going to Namche, which has the only post office in all of Khumbu, Gerard dashed off a letter to one of his women and gave it to Beg Bahadur to post.

Returning up the hill, we met Wade and Paulette who

were looking for a garden in which to pitch their tent and a teahouse to serve them food as they had no kitchen with them. They were on their way to Gokyo for a short trekking holiday. Without asking Gyalzen's advice, I suggested they camp with us and share our kitchen. But Gyalzen, when he arrived, was not pleased. His chest infection was still bothering him; he looked haggard and was breathing heavily. I supposed he saw the presence of the Pheriche doctor as a conspiracy to undo the good incurred by the lama's prayers and his inhalations of the night before.

'How many people?' Gyalzen asked.

'Just two,' I told him.

He had arranged to pitch our camp by a mani stone in the walled field of a cousin, like Gyalzen a member of the Paldorje clan, and one of the headmen of the village. Although the cousin was away, his wife said we could share their fire and eat with them in the long room of their well-kept house.

'Okay,' Gyalzen said, but from the tone it was clear that Father had been relegated to a chicken coop for the night. At dinner I could make out enough of the conversation to know that Gyalzen was complaining about me to the mistress of the house. Gerard was oblivious of this minor drama, serving Wade, Paulette and Uncle Pasang an aperitif of Ricard. He was happier, having found anisette-drinking companions (Gyalzen wouldn't touch the stuff because he said it gave him a headache; but Pasang literally lapped it up, making him more human in Gerard's eyes). Wade and Paulette also expressed interest in learning to play tarot, a French card game with certain similarities to bridge. Paulette spoke some French, and Wade was willing to try, which in Gerard's mind gave them much *sonam.*

Kanche looked over at me between one of Gyalzen's tirades and laughed. Gyalzen is not very tall, about one metre seventy (5 ft 8 inches). Kanche, a strong, handsome woman, was almost six centimetres (2 inches) taller than her husband and certainly as strong. She is an earthy person with a ribald sense of humour and a character forged from iron.

'Gyalzen mad with Father?' I asked her.

'Gyalzen very mad,' she answered, laughing again, as if to say, ignore him, he'll get over it.

Before supper of soup and sharpa, Wade told us he had climbed Mount McKinley in Alaska with an Australian friend, Gary Scott, who sometimes worked as a trekking guide. Both Wade and Paulette thought highly of Gary, who confided to them that three years before he had encountered a yeti when leading a trekking group up Chhukhung valley. They had stopped at Bibre along the way for lunch. Bibre is a summer yak-herding settlement of not more than a dozen rock houses situated at an altitude of 4,550 metres (14,947 ft). Two of the group strayed from the main party and after a while Gary went looking for them. Walking over the shoulder of a ridge he saw them standing dead still, staring at a point about 300 metres (985 ft) away. As Gary turned to look at what they were starting at, he said he felt the hair on the back of his neck stand on end. To his right at some distance were two yetis moving about the scrub and rocks looking for something.

One was an adult, he said, about the size of a small man or woman; the other was shorter and seemed to be a yeti-child. They were reddish-brown in colour, with long hair over most of their bodies and they blended perfectly with the countryside. When they saw Gary, they became agitated. The bigger one turned and picked up a rock with two hands and threw it in Gary's direction, then both animals ran off and he lost them from view. Gary told Wade he had never mentioned the encounter to anyone, fearing he would be laughed at.

We played three hands of tarot and around 9 p.m. retired to our tents. That night the BBC reported that Mrs Thatcher had warned Parliament that an economic recession was stalking the world. We were hoping for snow. The possibility of worldwide recession was something we had not considered.

Chapter 8

Donag Tsho

By every measure yaks are noble-looking creatures. They are rugged and shaggy, with horns that curve upward and outward. They have adorable ears, a fine coat and long, elegant leggings. They make you want to nuzzle up to them. But that can be most inadvisable, even dangerous. For in spite of their appearance, yaks are endowed with mean characters. They can be downright nasty, unappreciative and unpredictable, not to say stupid. They can stampede without warning, and when they lose their heads they stop for nothing: certainly not for people and sometimes not even for cliffs. When one lowers its head and starts pawing the turf, you know it's time to head the other way. Even Sherpas, who have been living with yaks for several hundred years, pay them the greatest respect when their moods become sombre or fickle. I have watched a crowd of Sherpas scatter across Tengboche common when four yaks, for no apparent reason, decided it was time to bolt. And yet on other occasions they let themselves be overloaded, scolded, beaten with sticks, stoned and sworn at without showing the least emotion.

Marsan and Tsao were no exceptions. Marsan, with a grey coat and leggings, was the more senior of the team, being twenty years old. If yaks have been well fed and not overburdened during their working years, Gyalzen claimed they can live to twenty-seven. Tsao, who was black and white, at fourteen years of age was therefore in his prime. Both were males; cow yaks are called naks.

Although Gyalzen frequently drove Marsan and Tsao, they definitely belonged to Kanche. They were gentler with her than with anyone else and would perform feats of endurance for her that no other person could command or coax from them. Of course she was the one who usually fed them and this certainly helped, but it is said that yaks are often more tolerant with women and indeed I have seen trains of them being driven by the slenderest of Sherpanis.

Gyalzen perhaps knew more about yak psychology and was a better teacher when imparting yak-handling knowledge to his children. He was in any event a kind person with animals, something that most Sherpas, while they would never kill one, are not. An animal, after all, is a low form of reincarnation, that is to say someone who had fallen low in a previous life, committed some grave act of *dipka*, or generally had been a miserable son-of-a-bitch and therefore, reincarnated as an animal, was to be treated as such.

Whenever Pasang Tshering or Kunga Pemba travelled with us, Gyalzen always reminded them not to harry Marsan and Tsao when they stopped to pee. He also taught them that yaks should be allowed sufficient time to drink at streams, and never to yell at them when the trail became perilous but let them find their own sure-footed pace. A yak, after all, represents a considerable investment for a Sherpa and, also, it is more than just a beast of burden. From yak hair, Sherpas make wool, from nak milk comes the greatest of all Sherpa delicacies, rancid butter, and, provided someone else has done the slaughtering, they tan its hide and eat the meat.

Yaks, I might add, most of them at least, are not as nasty with children. Of course it helped if they knew the children, for in general they don't like people whom they don't know. And they showed their dislike to anyone who came too close by suddenly lifting their heads with a nasty twist, scything the air with one of their horns.

I had tried to bribe Marsan and Tsao into an acknowledgment of friendship by offering them food. But yaks, I learned, are creatures of principle, which perhaps contributes to their aura of nobility. If they don't know you, they won't deal with you. In fact they'll try to bunt you away or, if really angry, impale you with that sudden upward and sideways thrust. So in the end I decided it was better to give them wide berth until they came to accept me as one of the family. In the meantime, the biggest mistake would have been to show them fear, because they can sense it and thereafter never offer the least courtesy. At all times it was important to remain impassive if one wished to retain the least semblance of an upper hand. Besides being empty-headed, yaks have cold and miserable hearts. This was why, I concluded, they were yaks and not somebody else.

On my first trip to Gokyo I remembered meeting three yaks in the command of a colourful Sherpa. They were on their way downhill, unladen. The trail fell sharply to the Dudh Kosi. Unconcerned, I stopped to watch them pass, standing on an outside boulder.

'Never,' the Sherpa said, stopping to lecture me, 'stand on the outside when a yak passes. Always remain on the inside.'

This was wise advice, a lesson I have remembered ever since. I have subsequently heard stories of tourists and even Sherpas being knocked over the precipice by a yak not realizing the width of its load or perhaps having been nudged sideways by another yak attempting to get to the head of the line. I decided there and then this would not happen to me.

Another yak idiosyncrasy is its way of walking, with head down, as if sniffing the trail ahead of it. This is contrary to the buffalo, and yet yaks and buffalos are said to be cousins. The Nepalis, every bit the equal of the Sherpas as story tellers, have a tale about this. It seems that at the beginning of time the yak and buffalo were the closest of friends. They lived on the gentler slopes of the middle mountains together, until one day they had a row and vowed never to see each other again. And so the buffalo went down into the plains and the yak climbed higher into the mountains. But with the passage of time they missed each other's company and wished they had never parted. Now when a buffalo walks through the fields of the plain he keeps his head upward, hoping to catch sight of his friend in the mountains. Likewise, a yak walks with his head down, looking into the valley, hoping one day he will see the buffalo again.

Next morning Marsan and Tsao decided to be ornery. Not even Kanche could get them to stand still long enough for her to place the wooden saddles on their backs and sling the girths under their middles. Gyalzen and Pasang had to tie ropes around their horns and hold them until they submitted to being laden with expedition drums, tents, climbing gear and the rest of our baggage. Now the rule of thumb is that a yak can carry roughly twice as much as man. The standard load for a Sherpa of any size, age or sex is 30 kilograms. Altogether, between the yaks, the three teenaged Sherpanis, Gerard, myself, Gyalzen and Pasang we were transporting 260 kilograms, and this for a two-week outing. The only person who was not carrying cargo was Kanche, a nursing mother, because she had Mingma Ramu on her back in a wicker basket.

Our destination that day was Nah, a hamlet of a dozen yak herders' stone huts, two of them transformed into primitive lodges that were abandoned in winter. Situated at the foot of Ngozumba Glacier's terminal moraine, Nah was 10 kilometres away (6.25 miles), but its altitude of 4,370 metres (14,355 ft) meant a net gain for the day of 530 metres (1,740 ft), except that the trail was uneven, constantly rising and descending, so in reality we probably climbed three times that much. We said goodbye to Wade and Paulette, as they were taking the west-bank trail to Gokyo via Dolle and Macherma. We climbed to Konar, another yak-herding settlement belonging to the people of Phortse, in less than an hour. Konar looked like a Scottish crofters' village. Gyalzen said that yetis had been seen on the lip of the plateau above Konar in wintertime. He added that once a yeti had even tried to descend to Konar after a heavy snowfall but when the herdspeople saw it they started yelling, shouting, and banging pots and pans. They kicked up such a din that the yeti turned around and climbed back to its perch. Pasang, who had the greatest respect for yetis, confirmed that during the monsoon the smaller pointed-head variety, the mitey, often frequented a small lake up there.

After Konar the trail climbed to a spur at 4,278 metres (14,053 ft), on which there was a chorten. Caves dotted the rocks below it and we saw two tahr ambling about the ledges. Our direction was more or less north, with a slight tilt to the west, but until reaching the chorten we were on a south-facing slope, without snow. The moment we crossed the spur behind the chorten a thick cover of hard snow blanketed the north side, in reality a near-vertical corridor that descended 400 metres (1,300 ft) to the Dudh Kosi with the abruptness of a toboggan chute. The trail snaked into the corridor, dropping 20 metres till reaching the backwall, then curved left and climbed again to a southern, snowless exposure. Gerard, ahead of me, called up to ask if we should cut steps with our ice axes. Gyalzen and Pasang had on reasonable boots, but Kanche, with the baby on her back, and the three Sherpanis were wearing tennis shoes.

'What about it?' I asked Gyalzen.

'No need' he replied. 'We go *bistaari, bistaari.*'

Once Gerard and I were in the shadows of the corridor it became evident how treacherous this passage really was. Even with climbing boots on I felt a rush of adrenalin shoot through me, and took it very, very slowly. But what would the Sherpanis, with their heavier loads, do? Gerard, the first to reach a secure place, unbuckled his ice axe and started chipping footholds out of the ice and compact snow.

Until then, the walk had been agreeable. Now we knew that on the north side of every spur lurked danger. We walked ahead as quickly as possible to the next spur to cut new steps. This was exhausting work. As Gerard was more proficient than I, he left me the task of stopping the yaks until he had completed a safe passage. Gerard, however, is a perfectionist and was not satisfied until he had carved a virtual staircase. But the yaks, while surely appreciating the steps, weren't interested in waiting around.

The first time, I stretched out my arms and stood in the middle of the metre-wide trail so they couldn't get around me. Marsan and Tsao were so astonished that they stopped. The second time, Marsan, who was in tail position, kept on coming, almost pushing Tsao over the precipice.

'*Dégage, Gerard,*' I yelled.

He read the urgency in my voice as panic — and he wasn't far off. But then he had never seen yaks lose their heads for no reason.

'You hit them across the horns with the handle of your ice axe. It reverberates in their skulls and brings them to reason,' he told me.

'To hell with you. I'm moving on,' I said.

The last shoulder before our lunch stop at Thore proved almost fatal. Kanche fell, but was already at the bottom of the icy pitch and didn't risk sliding down the chute to the rocks below. But the Sherpani carrying the ski bag stumbled in the traverse and started slipping along the crusted snow. She managed to stop, then couldn't move for fear of slithering over the edge. Pasang, who was nearest, inched down and grabbed her by the hair, steadying her until she could get her feet back under her, then guiding her to safety.

We reached Nah in mid-afternoon and set up camp in a pasture between two houses. High clouds obscured Cho Oyu, Tibet was overcast, and the wind was chill, whistling in from the west. Partridges chirped and waddled along the stone walls, watching us. Gyalzen seemed in better spirits but his lungs were, if anything, worse. He bought hay from a local yak-herder who had not yet descended to Phortse for the winter. Above Nah, Gyalzen explained, Marsan and Tsao would have nothing to eat.

Over dinner in a rock shelter, Gerard ventured in his best English that he needed a Sherpani to keep him warm for the night. Kanche studied him for a moment before suggesting that he could have all three. That was too many, Gerard replied, adding that he would be too tired in the morning. Everyone laughed; our threshold for humour was low.

The yak-herder across the way had a good supply of raksi and sold us a full bottle. As we sipped it after dinner, Gyalzen launched into a round of story-telling. Hindus believe that on the night of the August full moon the waters of Gokyo Tsho possess sacred fertility powers. Childless couples come from around the world — Gyalzen's definition of the world was limited — to bathe in its glacial waters, after which they cover their eyes and pick up a stone from the bottom of the lake. Gyalzen put a hand across his eyes and mimed someone staggering into water. If the stone was black, they would be blessed with a boy. If white, they were promised a girl.

The yak-herders of Gokyo were bemused by this rite. Firstly, Sherpas detest, or perhaps fear is a better word, any water not contained in a bucket. Swimming is a sport unknown to them and they only have the vaguest notions about bathing. Wading nearly naked into a glacial lake was in their mind clearly senseless. Moreover few of the pilgrims were used to the rigours of high-altitude trekking or the cold climate at glacier junction. They literally put their health at risk for a chance of having a child, which was touching perhaps, but, from the point of view of the herders who had spent their summers at Gokyo since first being carried there on their mother's back, such efforts were patently futile.

In the morning my leather climbing boots were frozen. After dismantling our tents and folding them away, it was minus 10 degrees centigrade and rather than wait for the yaks to be loaded, Gerard and I set out across the Dudh Kosi and began the climb to Gokyo. At the first of the Dudh Pokhri lakes, four sheldrakes, their golden plumage reflecting off the water of a still unfrozen bay, noted out passage. The snow cover had become continuous, but the trail was well-trodden, marked with toilet paper and pill separators; we made Gokyo in two and a half hours.

At an altitude of 4,750 metres (15,585 ft), Gokyo is only 57 metres (187 ft) lower than Mont Blanc, highest of the Alps. It was overflowing with trekkers, including a group of twenty doctors from Massachusetts. The lake was milky green and it was the first time I had seen it unfrozen. We lunched there in the sun and left soon after for Donag, walking up an ablation valley between Gokyo Peak on our left and the lateral moraine of Ngozumba Glacier. While Gerard climbed to the top of the main moraine, I made for the family of stones – a series of big and little cairns on top of a ridge that from a distance looked like travellers in the scree. At one point I passed a pile of partridge feathers, the remains of somebody's dinner, but as the snow was hard the predator had left no tracks. We were no longer on a trail, as nobody had ventured north of Gokyo since the last snowfall, and this pleased me.

Gerard continued along the top of the moraine, 80 metres above me, while I headed for Donag Tsho's shoreline, following the surest way I knew of finding the *yersin*, which was hidden in a trough between three moranic ridges. From the moraine, Gerard could survey the glacier on his right, which was not visible from the ablation valley, and on his left he had an overview of the moranic ridges. This enabled him to see the depression where the *yersin* lay hidden. The glacier, covered in gravel, at that point was more than a kilometre wide. From its headwall under the Cho Oyu-Gyachung Kang arc, the river of ice and gravel extended 20 kilometres (12 miles) southward, making it one-quarter longer than Mont Blanc's Mer de Glace and a third again as wide. All that I could see from my lateral trough was the northern arc and the twin peaks of Kangchung rising from the glacier's eastern moraine like the giant superstructure of a ship immobilized in a sea of ice.

When finally I arrived within sight of the *yersin*, Gerard was already sitting on the crest of the biggest hill opposite it, waiting. He was impressed by the cold. So was I. Within minutes of sundown, the mercury had plunged by 30 degrees centigrade.

The *yersin* was about 100 metres from Donag lake and maybe 80 metres above it. A maze of animal tracks criss-crossed in and around it.

'I'm not going to spend the night in that ice box,' Gerard called from his hilltop.

By this time Pasang had caught up with me and said the yaks were stumbling badly and risked cutting their hocks on the crusted snow. He was doubtful they could reach the *yersin.* Gerard continued to argue for a campsite on high ground, citing the 'inversion of temperatures,' which is part of mountain physics: heat rises, so that hilltops are often warmer in winter than the valleys that surround them. This phenomenon I knew worked in the Alps, but I was uncertain of it being applicable in the Himalayas. In any event we were in rocky terrain where water does not remain on the surface but sinks into underground channels. And we needed a ready supply of water. At least at the *yersin* we had snow to melt, while on Gerard's hilltop the snow had been swept away by the wind. We were now at 4,900 metres (16,076 ft), higher than Mont Blanc, and I was not going to carry water from the lake up to Gerard's mountain top. But he remained unimpressed by my argument.

Pasang and I walked back to help the Sherpanis and encourage the yaks. Marsan had cut his leg. Kanche had no gloves and her hands, holding the tumpline attached to Mingma Ramu's basket, were blue with cold. I gave her my gloves as I had another pair in my pack. We got the yaks moving again. Gerard was still complaining about the deep-freeze effect and insisted we go higher. But Gyalzen and Pasang opted for a middle solution, designating a turf-covered knoll that stuck out of the snow with a large rock on its flank as our campsite for that night only. In the morning we would choose something better.

Gyalzen and Pasang started assembling the kitchen. Then, while Kanche breast-fed the baby, they set up a single tent for the seven of them next to the kitchen. Gerard and I levelled platforms in the snow for our tents, downhill from the kitchen. It was pitch dark: no moon, but the sky was carpeted with stars. By 8:30 p.m., dinner was over and we rolled into our sleeping bags.

About 10:30 p.m. I was awakened by wind whistling through the tent. There was not much room in the sleeping bag as I had my boots and snow gaiters inside to stop them from freezing and I lay perfectly still, listening to the silence. At first the sound was faint, but it grew stronger: a tsonk-tsonk-tsonk that repeated itself several times, moving from north-west to north-east in the rocks above our campsite. Silence for a few seconds, and then there was an urgent, answering tsonk-tsonk from behind the kitchen rock. This tsonk had a different tone, and was nearer. Minutes later I heard a final tsonk-tsonk-tsonk, this time from the south-east, between our campsite and the *yersin.* After that, only the silence of the wind.

Chapter 9

Yeti Country

So this was yeti country! I struggled in the restricted confines of the tent to put on my boots and down jacket. It was cold, cold, cold, with frozen condensation clinging to the roof. At 7 a.m. the sun was not yet up, but certain I would encounter a wealth of yeti tracks left by the previous night's visitors, I crawled outside into a minus 20 degrees centigrade ice locker. The snow, already a month old, was so hard that no fresh tracks were apparent near the tents. Scores of old ones — fox, martin, hare, and something that Pasang called a mountain cat — confirmed that even in this glacial environment there was a resident population that by the look of things enjoyed a flourishing social life.

Disappointed, I climbed to the crest of the monticule behind our camp. The countryside was like a series of rubble heaps that had been bulldozed into place by Ngozumba Glacier in one of its earlier phases of progression or regression. This giant 'earth' mover is one of the very largest glaciers in these latitudes, stretching, snout to backwall, a distance equivalent to the full length of Manhattan Island, or from Hyde Park Corner to Heathrow. For most of this distance it is a stony-road glacier, covered by detritus, including boulders as big as houses, and carrying a full load of scree, gravel and sand. No vegetation lies on the glacier, of course, nor for the most part on its moraines and monticules. Some scraggly juniper and rhododendron bushes less than 40 centimetres high cling to the southern exposures, where tufts of wiry grass and occasional brown moss offer meagre pickings for a yak.

I was working alone, as everyone else was still sleeping, and therefore could afford to take my time. Between our camp and Donag Tsho, beyond the top of the first monticule, I came across another mound of partridge feathers by a boulder the size of a minibus with an overhang large enough for a fox or wolf to hide under. I knew wolves existed in the area because Gyalzen and I had followed one's tracks when we climbed Renjo Pass from Gokyo eighteen months before.

I continued to the next large boulder, three metres away. In its lee was a series of imprints that looked strikingly as if they had been made by a primate. Only one in the series was protected from sun and wind, and consequently was not too deformed. It had five barely discernible toes, was about a size-nine foot, a little smaller though broader than my own without the boot, and who knows how old: a week, a month?

Though this footprint resembled the one in Shipton's 1951 photograph, which has become the international standard for a yeti imprint, it was certainly nowhere near as sharp. And while the footprints observed by Shipton on Menlungtse Glacier, a mere 25 kilometres (15 miles) to the west across nothing but barren tundra, suggested a biped with a metre-long stride, the ones on the Donag monticule were close in line, only centimetres apart. This puzzled me for some time, until it dawned on me that the animal was creeping around the rock, possibly stalking something, in any event pacing itself slowly until it could see what was on the other side. But still I was more disappointed than excited, because I had expected to find fresh tracks and these were weeks' old. My inclination was to disregard them. They offered little useful information, except that I had now convinced myself they were bipedal, and not made by a bear. I followed them for a few metres, but once they went across an exposed basin, where no shade protected them, they became blurred and distorted, eventually losing themselves in the rocks on the south side of the monticule.

I had considered Donag Tsho to be likely yeti country for several reasons. First, as the absence of human tracks on our way up the ablation valley seemed to confirm, it was isolated. Second, it had a modest history of yeti sightings. Nothing prolific, like the hamlet of Macherma further to the south, but Donag was definitely on the map of yeti localities. Donag's *yersin — yar* means summer in Sherpa-ka and *tsin* is a temporary settlement — belonged to people from Khumjung, above Namche. The husband of the didi who ran the Kalapattar View Lodge in Gokyo was from Khumjung and she had told us that a few monsoons back a yeti had been seen from a distance of several hundred metres strolling along the top of the moraine above the *yersin*.

What intrigued me even more was the account of a 1954 visit given by Ralph Izzard, leader of the first expedition to set as its goal the finding of a yeti. Although this expedition was generally regarded as a failure, it actually accomplished a good deal.

In 1954, the territory was unknown and the hamlet of Gokyo was abandoned in winter. So, for that matter, were Nah and Macherma, and their stone huts had not yet been converted into trekking lodges. Izzard made his first sortie to Nah at the end of February 1954 and discovered two sets of footprints there. The footprints headed south towards Macherma but became lost in terrain where the snow cover had melted. With Gerard Russell, an American naturalist, Izzard followed the footprints northward, hoping to find where they originated. At the first of the Dudh Pokhri lakes above Nah, where we had seen the sheldrakes, they discovered a 'positive maze of yeti tracks'. It took them some time to interpret these tracks and figure out what had happened.

It finally occurred to them that two animals had converged at the lakeside, one coming across the glacier from Dragnag, at the foot of Tshola Pass, the other descending from Gokyo Tsho, the third lake in the string of five. The animals had arrived at the lake more or less simultaneously and spent some time circling each other. Izzard and Russell gathered from the depth of the imprints that one animal was lighter, possibly a female, the other a heavier male, or possibly an adult, who teamed up with a less mature companion, for they continued together towards Macherma.

Izzard and Russell followed the set of tracks that arrived from the north. They discovered that this animal had skirted Gokyo, descending from Renjo Pass, above the eastern end of the lake, which meant it had come from Nangpa valley.

Izzard, in his book *Abominable Snowman Adventure* (Hodder & Stoughton, London, 1955), wrote: 'Although somewhat spoiled by melting and wind drift, the majority of the prints showed a clear impression of one big toe and at least three smaller ones . . . We judged them to be eight to nine inches [20 to 23 centimetres] long and possibly four to five inches [10 to 13 centimetres] across and the length of the stride was two feet, three inches [67.5 centimetres] long. Our general impression was that although smaller they otherwise corresponded exactly to those photographed by Eric Shipton . . .'

Accompanied by their Sherpa guides and porters, Izzard and Russell reached Donag Tsho, the fourth lake in the chain, two days later and found another 'maze' of tracks. 'The ground literally abounded with yeti tracks,' Izzard wrote. They had picked up the first of these tracks — belonging to a third yeti — north of Gokyo and followed them until they tailed out above Donag. They camped two nights at Donag, decided that the abundance of tracks on the frozen lake was due to the presence of a fourth yeti, but they never discovered where it came from or where it went. One of their Sherpas gathered a bagful of supposed yeti droppings, said to have been more or less human in size and texture. As it snowed on their second night at Donag, they returned to base camp under Tengboche ridge where they dissected their scatological find. Other than identifying tufts of mouse hair, they learned nothing new from the droppings about the dietary habits of their quarry.

We gathered for breakfast as the sun came over the eastern ridge. Kanche, one breast out in the cold nursing

Mingma Ramu, was laughing. Gerard was at his lowest, suffering badly from a combination of cold, lack of sleep due to Cheyne-Stokes breathing — a high-altitude syndrome that wakes you during the night gasping for breath — and a persistent headache. Nobody but Father had heard the tsonking of the night-time visitors. After breakfast of tea and tsampa, Kanche, the three Sherpani porters and the yaks left for Pangpoche.

As the battle for the campsite immediately resumed, I forgot about the tracks atop the monticule. Gerard still wanted to move our tents uphill. Gyalzen and Pasang favoured a site by the lake. I wanted a comfortable observation point somewhere in between, with a source of water nearby. Gyalzen won, largely by taking matters into his own hands. The lake possessed a broad and rocky beach that resembled, except for the water, a wintry lunarscape, totally void of plant life. Near the top end of the beach was a large boulder. Gyalzen, son of a Sherpa 'engineer', could already visualize the community he would build there. The boulder became the kitchen's north wall. He and Pasang would erect a dry stone wall for the south end of the kitchen, string a rope between the boulder and wall as a central beam and drape our green nylon sheet across it for a roof. He planned to pitch one tent for himself and Pasang three metres to the north of the kitchen and level the sandy point ten metres to the south for Gerard's and my tents, centimetres from the lapping water.

'But Gyalzen, what if the level of the lake suddenly rises?' I asked. This can and does happen, caused by glacier rupture or a large rockfall.

Gyalzen laughed. 'Water going down, Father. Lake will be lower each day,' he said.

Donag Tsho is, to my mind, more beautiful, and more rugged, than Lake Louise in the Canadian Rockies. It is 1.6 kilometres long, has a maximum width of 700 metres and lies in a glacial basin between Ngozumba's western moraine and the main north-south spine that separates the Dudh Kosi or Gokyo valley from Nangpa valley to the west. Its southern shore is a line of black cliffs that taper into elegant spires, some more than 5,500 metres (18,400 ft) high. From its northern shore rises a steep rounded hump, baptized North Slope, that was not quite as high as the spires. North Slope was covered with brown grass, and in places stunted juniper and rhododendron bushes. During monsoon, yaks pastured on it. Other than the sparse grass and stunted bushes, there was no other vegetation, the nearest tree being at Dolle, 17 kilometres (10.6 miles) to the south.

At the lake's western end a small stream flowed in from the scree valley above, but no stream flowed out. And yet, each day at this time of year, as Gyalzen had pointed out, the water level dropped a few centimeters. The lake really sat in a crater, with a ridge at the southern end barring any exit for a stream. There was, however, an irregular gurgling sound that resounded off the southern cliffs. This noise was caused by a natural siphon that drew the lake water into a subterranean channel. The water broke surface again two-thirds of the way down the ablation valley to trickle into Gokyo Lake.

I remained unconvinced that the lake might not suddenly change level and pointed instead to a brown, black and orange-speckled rock about 20 metres above the lakeshore. This rock had already served as a campsite, for someone had erected a stone wall under an overhang facing the lake. This, I thought, would make a good storeroom and observation blind.

'Yes, already seen it,' Gyalzen said. 'But here is better.'

Gyalzen was walking slowly that morning and it seemed that each step was an effort for him. He had already decided that, come Wednesday, he would return to Pangpoche so that he could go with Kanche to Saturday market in Namche. My decision, I told him, was to store the climbing and observation equipment under the rock; I would place my tent on a snow platform in front of it. Gerard could decide where he wanted his tent. Gyalzen was disheartened by my decision, as if his judgment had been placed in doubt. He turned sullen.

Gerard was not at all happy with this solution either. He

assumed that I had been railroaded into it by Gyalzen. 'It seems we no longer have the right to decide anything,' he complained.

Our first day at Donag was consumed with moving into the new campsite. While Gyalzen and Pasang constructed a wall for the kitchen, Gerard and I erected our tents. Gerard selected a southern exposure hard against the rock, as opposed to my western-facing one. The rock was the size of a chalet and became known as Rock House. While leveling the ground for his tent, Gerard uncovered some scat and when later we showed it to Pasang he said simply, 'Mountain cat.' We had no idea what he meant by that, but thought at first it might have been a lynx. Unfortunately I did not think to keep the scat for analysis and later I was told no lynx lived that high, but then again neither does the yeti, since, according to science, the yeti does not exist at all.

No sooner were we installed than the sun sank behind Gokyo ridge. It was 3:35 p.m. and without sun the temperature immediately plummeted from a balmy 18 degrees centigrade to well below zero, on its way by nightfall to around minus 25 degrees centigrade. Half an hour after losing the sun we gained the company of Wade and Paulette, and Gerard's spirits rose. The kerosene lamp we had brought from Pangpoche I suspected had seen service with the Gurka regiments in the First World War. While Wade and Paulette set up their tent on the sandy point, Gerard fiendishly attempted to repair the lamp so that we could play tarot after dinner. Once his tinkering was complete he wanted Gyalzen to fill it with kerosene.

'Tomorrow,' Gyalzen told him. These two were not made for each other.

Finally, as we finished our noodle soup, Gyalzen consented to test the lamp and a burst of flame almost lost us the kitchen roof. The lamp was not repairable. Gyalzen said he would toss it into the Dudh Kosi on the way home. The lake was sacred, though. Both he and Pasang had lectured us not to wash or throw dirty water in the lake so as to avoid angering the *klu* who lived there. No garbage could be burned, either, in case the bad odours upset the gods. The gods didn't seem to mind if we buried our trash on the beach, however, and so we made a waste pit in the stones, to be covered over when we left.

We played two hands of tarot by the light of our head-lamps, sitting on cold stones in the smokey kitchen, with gloves on, and a cool breeze wafting in off the lake at kidney level. I was not really concentrating and only agreed to play a third hand for the sake of Gerard's morale.

Next morning Pasang built an altar of three stones on a flat rock by the kitchen and burnt some rhododendron shoots to appease the gods. This was to become a daily ritual. The shoots gave off a fragrant odour that the gods were said to like. Gerard decided before breakfast to wash some clothes and asked Gyalzen for hot water and a basin. Gyalzen was unwilling. We had a single butane burner and a small fire of juniper branches, and Gyalzen needed all this heating power to make tea, cook eggs and warm the milk for breakfast. Gerard didn't understand and thought that Gyalzen was being deliberately difficult with him.

While we ate breakfast, I set up the spotting scope and studied some tracks at the far end of the lake. I thought they looked promising and wanted to investigate them. We said goodbye to Wade and Paulette — they were returning to Pheriche — and set out, I by the lakeshore and Gerard by North Slope. When I got to the stream, I found the tracks too warped by the sun to identify and began climbing the dorsal moraine between what once had been the beds of two glaciers, to find Gerard waiting for me a little further up.

We continued towards the base of a mountain that, from our campsite, was hidden by the shoulder of North Slope. Gerard maintained a steady pace up the bare mountain to a snowy crest at 5,200 metres (17,060 ft). I followed more slowly, stopping every few metres to suck air into my lungs. Until the ridge we had been in short sleeves. It was now 1:30 p.m. and on the saddle we were treated to a view of Everest's north face. But the moment we stopped, a breeze stirred and it was time to put on a sweater.

'We could have made it to the top if those two idiots would only decide to move a bit faster in the mornings. Your organization is no good,' Gerard complained. 'We will never accomplish anything at this rate.'

This I thought unreasonable. It was our first operational day at Donag and we had not yet established a routine. Anyway I told Gerard I objected to his calling Gyalzen and Pasang idiots.

I put on my pack and started climbing through snow that in places was up to my thighs and I found it exhausting even though the slope was not steep. I was mad. I didn't want Gerard undermining my relations with Gyalzen. But that was the narrow-minded side of Gerard I had never encountered before, although we frequently ski-toured together in Chamonix.

The trouble, I surmised, was that Gerard's women spoilt him mercilessly. Previously divorced, he seemed to have an ideal relationship with Josiane, who was two years older and would do anything for him, including share his heart with a younger woman, though for how long was not clear. Although they had actually shared this kind of triangular relationship once before, as far as I could gather Josiane knew nothing of Laurence. But even if she did suspect there was someone else, Gerard was supremely confident he could keep both women happy.

'Don't delude yourself,' I had told him. 'It'll never work. Anyway, you should stick with Josiane. She knows you best. And besides, you've already told me she cooks the best.'

The climb to the 5,400-metre (17,716-ft) summit had been no more challenging than ascending a kilometre-long garage ramp; we were now suspended in the middle of a circular valley with a 300-metre drop-off on three sides and a 360-degree view of the mountains around us, all of them much taller, from Gauri Shanker in the west to Makalu in the east. At our feet was a carpet of snow, ice and scree, forming two kidney-shaped basins that touched top and bottom with us above and between them.

The western basin contained two frozen lakes not shown on the map and a confusion of tracks, all heading for or returning from five or six large boulders or rock overhangs. Some of the tracks appeared large, while others must have been made by a fox, though from our distant niche it was impossible to tell. Gerard was interested in a double set leading up the west shoulder of North Slope, another kilometre away. Although they looked really sharp, we could not decipher where they had come from and precisely where behind the backside of North slope they were heading. We were tempted to cross over to them, but as we had expended more then three hours getting to our superb vantage point it was too late to descend into the basin, climb behind North Slope and return the 3.5 kilometres to camp — a distance, say, from Buckingham Palace to the Bank of England — before nightfall. Regretfully we left them for another day, and headed back to that bit of lunarscape we now called 'Coconut Beach'.

At supper that night Gerard had another headache. Neither of us had much appetite and afterwards I went to my tent to prepare for my first early-morning vigil with Orion, the electronic nightscope.

Four a.m. No moon, dark sky and the dipper was upside down. At this altitude, just to survive was an effort. Crouched in my tent I struggled to put on my one-piece polar suit and was left panting.

I crawled out and moved slowly uphill to the shelter of some rocks. Every footstep seemed to echo in the crisp night air. I switched on Orion's batteries, listened for the whirring noise of the light intensifier building up power, and tested it. There was hardly enough luminosity for the intensifier to work. Disappointment. I stayed behind my pile of rocks for an hour. There was no wind but I was freezing. At 5:20 a.m. I returned to the tent, hardly able to move fingers or toes. My strategy had to be rethought.

Next day we decided to climb North Slope. Gyalzen was still not well enough to accompany us. He had been shivering in the sunlight at breakfast and this was a troubling sign. So we set out alone, climbing steeply for the first 200 metres (660 ft) from the lake onto a slightly convex plateau before continuing toward two well-rounded humps separated by a snowclad saddle. The right hump was higher and extended along North Ridge towards a series of multi-coloured gendarmes that overlooked Ngozumba Tsho, the fifth lake in the milky chain. Because they rose like a row of spindles on top of a sand heap I called them 'The Needles'. Gerard climbed directly to the right hump and I made for the saddle.

On the south side of the saddle the snow had been transformed by the sun, and was as hard as double-ply cardboard. I could make out some traces of tracks coming from the shadows of the far side, where the snow was deep and still powder. When I crossed north of the saddle the tracks came alive. I could see that they had come from a rocky ridge behind a frozen pond, and led to where I was standing, a distance of 180 metres. The bowl surrounding the frozen pond was as bare as a jar of cold cream — not a clump of turf in sight, only white tundra.

I descended alongside them for 10 metres (30 ft) and judged them to be a single set, suggestive of a biped with a small stride, not longer than my own. The loose snow made it impossible to tell how large the footprints were, except that by human standards they were not large, perhaps the size of a teenage boy. They were definitely coming from the frozen pond, up the 45-degree slope to the saddle, heading in the direction of Donag Tsho, maybe two kilometres away, and our campsite, that is to say from north to south. How old were they? A few days at a guess, though not more than a week seemed confirmed by a second look at them four days later, when I was better able to measure the rate of their decomposition. By the lack of other large tracks in the vicinity it seemed evident that this 'pond creature' had gone south.

After photographing the pond creature's tracks I joined Gerard on North Shoulder. We decided to eat lunch in the shelter of a large rock and gazing back towards the frozen pond we noticed sets of smaller tracks, made by a fox, perhaps, or some similar-sized animal, confirming that there was a surprisingly rich animal life around 5,500 metres (18,000 ft), even at the outset of winter.

As we huddled out of the wind, Gerard confided that he had been kept awake during the night by another bout of Cheyne-Stokes breathing. His head hurt and he was lovesick. He said it made him short-tempered and he feared he would drag down everyone's morale. He wanted to leave with Gyalzen in two days' time and return to France. I appreciated his frankness, and was relieved. I had found him demanding and coping with his moods had required too much of my attention. It was distracting me from the task of encouraging the yeti to disclose itself and I agreed it would be better if he did return home to sort out his life. I knew I needed the peace of mind.

I was disappointed that Gerard had shown no interest in Sherpa culture. We had discussed my total immersion theory before leaving Chamonix — of trying to adapt as much as possible to Sherpa life, eating their food and sharing their customs so as to blend into their culture and countryside — and Gerard had agreed with this approach then. But once at Pangpoche he quickly became disaffected, reasoning that the natives were unwilling to improve their lot. But most of all, his depression was due to the fact that he missed his women, his dog, his friends and the familiar surroundings of Chamonix. He seemed happier once he had announced his decision and we had talked about it.

The first ice appeared on the lake overnight and Tuesday, 24 November, dawned with no wind. Gerard did not want to leave Donag without trying our Yeti touring skis. Sixty or seventy metres opposite Rock House was a low ridge sloping down to the barren and untropical 'Coconut Beach'. The ridge displayed half a dozen sets of semi-obliterated tracks which so far we had neglected to investigate.

We put on the skis and went to inspect the nearest ones, with Gyalzen following on a pair of red plastic snowshoes. Halfway up the ridge we noted that this set divided in two.

We followed the stream that seemed the sharpest, leading to the lee of a rock the size of a news vendor's kiosk. In the shadow by the rock's downside, Gerard discovered four footprints that were reasonably intact; in fact two were almost perfect. Neither sun, nor wind, nor spindrift had deformed them — but they were the only two in the series that the elements had treated kindly. As the tracks continued beyond the rock they became blurred and were eventually washed out altogether. The two sharp imprints were comparable to, though smaller than, the 1951 Shipton footprints. We shared a moment of elation and then began to ponder what they meant.

And what can two footprints tell you?

Quite a lot, actually. To begin with, the foot that made them measured approximately 18 centimetres (7.1 inches) long by 12.5 centimetres (4.9 inches) wide. This compared to 32 centimetres (13 inches) by 20 centimetres (8 inches) for the Shipton footprints, but they were approximately the same size as those found nearby by Izzard and Russell. Our animal had one very large toe and a not too distinct array of four others. The tracks had come out of a gully and crossed the ridge diagonally, heading towards Gokyo. The gully separated the ridge from the monticule on whose summit four days before I had seen the first set of washed-out tracks. Coming out of the gully, the tracks showed a bit longer than normal stride of 65 centimetres (2 ft) but when the one animal approached the rock its stride slackened and it had placed one foot a few centimetres ahead of the other. At one point it crossed its left foot over the right.

These tracks had come in more or less a direct line from the monticule, 100 metres to the north. This led me to suppose the tracks on the monticule had been made by the same creature, though perhaps not on the same occasion. I was unable to decide what exactly could be inferred from this discovery, but I gave the creature to whom the tracks belonged the name of Monticule Man, as no doubt existed in my mind that he had been walking erect. They also indicated that Monticule Man was accompanied by a mate. This was a new element. I was unable to judge accurately when the tracks had been made or how heavy the animal might have been. Compared to the first tracks on the monticule it seemed possible that these were more recent, not more than ten days old.

So now we had three good leads — the double set of tracks seen from the ramp heading around the west side of North Slope, which we never followed up, the Pond Creature who in some way might have been connected with the North Slope tracks, and Monticule Man accompanied by a mate, both of whom could conceivably be connected with the still unexplained midnight tsonkers.

Gerard's morale was better after this discovery, but it was more because of his anticipated departure next day than the prospect of finding Monticule Man. He was sure mail would be awaiting him at Pangpoche from both of his women. Almost as an after-thought he mentioned that *again* during the night he had heard thumping from behind Rock House and thought that someone had shaken his tent. Other than the thumping, though, the animal (for he was convinced it had been an animal) made no other noise — no 'tsonking,' for example, such as I had heard on our first night by the *yersin.*

I was mystified as to why he had not mentioned this sooner. However it encouraged me for I, too, felt our presence was being observed. True, we had found no new evidence of Pond Creature's existence and nothing linked him with Monticule Man, though it was possible they were the same characters. But clearly *something* was out there.

In the kitchen tent that night Gerard and Uncle Pasang finished the bottle of Ricard. When I had met him at Lukla fifteen days before he had predicted he would leave when the bottle was finished. This proved no joke. He was full of advice, as he sipped the last of the Ricard, on how to attract whoever or whatever was observing our campsite. And on this point our views were perfectly in accord.

'There is no use moving about too much. You must stay in one place,' he counselled. 'Drawing the animal to you by

Evidence of deforestation and over-lopping of trees. This view shows the once heavily wooded south side of Lamjura Pass, between Kenja and Jumbesi.

Tengboche Monastery and its outbuildings covered by a fresh mantle of snow, with Ama Dablam in the background.

Gyalzen, Kanche and Uncle Pasang loading a reluctant Marsan, their lead yak, with the blue expedition barrels.

The first yeti tracks I found, on the monticule behind Donag Tsho. Only one print was protected from the sun and wind, and thus relatively undeformed.

Gyalzen and Kanche relaxing at the Shanjo campsite with Taboche Peak in the background.

Gerard Metral found this set of yeti tracks opposite Rock House, our Donag Tsho campsite.

The spoor of Shanjo's
wandering fox photographed
the morning after
the December snowstorm.

These are the same tracks,
doubled in size by sun
exposure. My ski glove
indicates the scale.

Gyalzen perched above Konar Pass.

Author in Sherpa dress for Lohsar with Kanche, holding Mingma Ramu, and Pemba Kunga.

Korean and American garbage at Everest Base Camp.

Nearing the top of Lho La with Pasang Tshering. Pumori is in the background.

Mount Everest viewed from the west.

putting out bait is the oldest trick in the world. It is the only one that works. First the birds will come. When the fox sees the birds circling, then he'll come to see why, and so on. It might take several days but you must be patient.'

Next morning, Wednesday, 25 November, Gerard and Gyalzen left; the birds returned, and Uncle Pasang was nervous about remaining alone on Donag's beach with an unloquacious Father. While he busied himself with improving the rock shelter so that I could use it as an observation blind, I revelled in once again being able to concentrate on the task at hand without feeling guilty about Gerard thinking he was isolated from the mainstream. The quietness of Donag was even more striking now that he had gone. I spent the day catching up on my notes, writing on a flat slab of granite beside Rock House. Pasang had supper ready at 5 p.m. and then it was back to my tent to prepare for my first night in the blind. Pasang was uneasy about being left as live bait at the lakeside. At sundown I thought I heard a fox barking at the far end of the lake. That seemed a promising omen.

By 6:20 p.m. it was completely dark. At 7:10 p.m. I took up position in the blind. Pasang had left me a small open window with a view of the kitchen, his tent and a full sweep of the lake. The door to my right permitted a lateral view onto North Slope.

Some sort of lunar phenomenon was occurring, perhaps an eclipse, because a shadow passed over the moon, obscuring its new crescent before it slid behind Gokyo Peak. I surveyed my territory with Orion. All was quiet on the beach. I switched on my headlamp for a moment to look for some pocket warmers, then quickly extinguished it. Several minutes passed before anything happened, then something bolted down a rock staircase less than 20 metres (65 ft) from Rock House. Rather, something moved the rocks as it fled down the staircase toward the lake. Whatever it was must have possessed reasonably good night vision, for I know I could not have negotiated that terrain without a headlamp, and certainly not at that speed. But apparently this animal could not see quite well enough to avoid boulders that moved, and, judging by the sound of those it displaced, it was a heavy animal, not light like a fox. I swept the area with Orion, but to no avail.

I was truly frightened, but amazed that I felt drawn outside the shelter in an attempt to follow the unseen creature. Between me and the rock staircase was a band of snow which I didn't want to cross in the dark for fear of ruining any tracks it might have left. For ten minutes I continued sweeping the area, but could see nothing moving along the lakeside nor climbing North Slope. I decided the best tactic was not to turn on my headlamp again to search for tracks, but to return to my window inside the blind and wait. I thought perhaps the animal would circle the monticule and, drawn by curiosity, come back. But there was no further sound, except for the crackling of the ice forming on the lake, which at times rustled like a flock of birds. Towards 8:30 p.m., I gave up and went back to my tent, nearly frozen.

By the light of my headlamp I wrote in my journal: 'Now I know the yeti exists!' Perhaps this was a case of hope triumphing over reason, perhaps not. For a few minutes I thought I heard a series of grunts coming from the snowy gully to the south of Rock House, near where Gerard had found Monticule Man's footsteps. But I could not be sure, and as I was already back in my sleeping bag I let them pass.

Chapter 10

Pasang's Discovery

The sun struck Rock House at 7:20 a.m. I dressed hurriedly and went out to see if I could find trace of the nocturnal visitor. I found nothing. I went down to the kitchen tent and asked Pasang if he had heard anything during the night. He said that around 10 p.m. some rocks had slid down North Slope and splashed into the lake. Interesting. Was this the same animal retreating to its home across the saddle by frozen pond?

I placed malt biscuits — all that remained for bait — on the tops of three boulders around Rock House. Then Pasang and I walked to the *yersin* to see if any new tracks were to be found there. On the way he explained that three Khumjung families owned the *yersin* and that they brought between sixty and one hundred yaks to pasture at Donag every monsoon. No new tracks were at the *yersin*, either. From the top of the moraine we surveyed Ngozumba Glacier. It creaked and groaned. Every now and then a piece of moraine broke off and, amid a cloud of dust, tumbled onto the glacier. To the south we saw three tourists walking up the moraine from Gokyo, so we returned to our hideaway, only to discover a couple walking along the lakeside toward us.

'Is this the highest lodge?' our visitors asked. They sounded Dutch.

'No, this is not a lodge,' I replied.

'Are you part of a mountaineering expedition?' the girl wanted to know.

'No,' I said.

'Just spending a few days here?' the man tried again, looking around. He had noticed Rock House.

'Yes.'

'You have skis?'

'Yes.' I was purposely being abrupt because I did not want them to linger with us, nor to tell their lodge-mates back in Gokyo that there was this interesting expedition up at Donag looking for the Abominable Snowman. I felt certain that had I told them the real purpose of our presence it would have brought an avalanche of abominable trekkers upon us. No thank you.

Not getting more information, they decided to return to Gokyo, abandoning their intention of continuing north, as I had told them Ngozumba Tsho was another three hours away.

For the rest of the day I was consumed by lethargy, the listlessness of altitude, and stayed close to my writing desk by Rock House. At 4:30 p.m. Pasang called me for dinner of noodle soup. He asked about my family. Pasang was normally quiet, but I sensed he wanted to talk. He was deeply religious and once had pointedly corrected me when I forgot in conversation to accord Sanga Dorje his proper title of Lama. It was like calling Cardinal Hume just Hume and he wanted me to show respect. He told me about his three children: Pasang Dawa, a year younger than Pasang Tshering; Pemba Yanchin, a girl aged seven; and another daughter called Hackma Yanchin who was about the same age as Mingma Ramu. Pasang's two eldest children went to school in the village. There was no school in December and January because it was too cold. He had one brother and three sisters, all married and living in Pangpoche. Both his parents were dead.

Pasang also said he had been five times to Everest's South Col, each time carrying a load and a half. With the proceeds from high-altitude portering he had purchased a herd of twelve yaks and had bred from them three calves. This meant he possessed a considerable capital, making him richer than Gyalzen.

I finally realized that Pasang was being talkative because he wanted to keep me down on the beach with him for as long as possible. He was not happy about spending the night alone while a nocturnal visitor roamed about. I told him I would be watching from Rock House, but he was not reassured.

Our nocturnal visitor was also preying on my mind. Did it live nearby? Had it left the human-looking stool on the rock above our camp? I had asked Gerard about it before he left. He said he was not responsible. That I should have figured out myself because of the absence of toilet paper. I did not have the courage to ask Pasang. Also Gary Scott's encounter with a rock-throwing yeti at Bibre had made me wonder. Until then, I assumed yetis were not aggressive, at least not towards humans, but I was no longer sure.

Orion worked beautifully. But there was no sign of life. Moon shadows rippled across the lake, which was also a lake of many noises, none of them the sounds of civilization. It was amazing to think that I had not heard a motor car since leaving Kathmandu, nor had I heard the sound of a jet. I was cold, so I left Rock House at 8:30 p.m. and went to bed.

Next morning the lake was very busy, moaning and groaning down in Siphon Bay, now covered with ice. Pasang wanted me to climb the west shoulder of North Slope with him. I said if no trekkers came into sight before 11 a.m. we would go. Sure enough, at 10:30 a.m. a single trekker appeared at the south end of the lake and crouched by the shore for so long that Pasang had to get the binoculars to make sure it was not a yeti.

The trekker finally got up and climbed to the top of the morain, showing no signs of having seen us. We zippered everything up and at 11:15 a.m. began to climb diagonally across North Slope. It was shirt-sleeve weather till we hit the snowline. Thousands of mouse holes dotted the brown turf and a flock of fifty or more *kaachas* fluttered overhead, some almost flying between us.'[1]

---

[1]My only translation for *kaacha* is 'small mountain bird.' We were at 5,000 metres when we saw them.

The snow cover was diminished since my last visit four days before. Pasang wanted to continue to the west hump, separated by the saddle from the main Needles North Shoulder. We got there at 12:30 a.m. and poor Pasang almost fainted at the sight of the footprints until quickly figuring out that Father had made them during his earlier visit to the saddle. The supposed yeti tracks had in the interval been almost obliterated and an absence of new imprints indicated that the nocturnal visitor had not headed for the frozen pond after his Rock House scampette.

Back at Coconut Beach nothing had been disturbed, not even the bait. Pasang came up to Rock House to listen to Radio Nepal in the last sun and — zip! — we both saw the tail of a small animal disappearing behind a rock. Zip! It moved to another rock. Soon it was scurrying from rock to rock up south ridge, leaving no tracks because its weight was not sufficient to break the snow crust.

'What is it?' I asked Pasang.

'Reymoo,' he said. 'Smells very good.' He meant that it had a keen sense of smell. 'Eats mice.'

It was a sleek and handsome ermine with light brown fur and a white tummy, about 40 centimeters long, and it moved like lightning across the hardened snow.

After an early supper I went back to Rock House and prepared for another vigil with Orion. There was a crescent moon and the wind rose in strong, unexpected gusts, lifting the kitchen roof like a great bellows and on one occasion knocking a tin cup off a rock shelf. That must have made Pasang miss a heartbeat.

Next morning, our eighth day at Donag, I offered our last slice of yak cheese to the ermine and set off on skis to explore Siphon Bay. I discovered that the black cliffs were honeycombed with niches inhabited by yellow-billed choughs. When I returned for lunch, feeling as lazy as could be, Pasang announced: 'Maybe tomorrow raksi come.'

I asked him to explain.

'Kanche promised to send raksi for me and Father,' he said.

Pasang placed extraordinary faith in Kanche. I had noticed that he worshipped the ground she walked on. She was an ideal: strong, energetic, earthy, five children. Pasang's wife, Nima Yanchin, was plump, short, liked fine clothes, was always combing her long silky-black hair, and was content with three children. The baby among Gyalzen's brothers and sisters, she was almost ten years younger than Kanche. I had the impression that she hounded Pasang to be somebody he clearly wasn't.

In the early afternoon two trekkers came down from the monticule to the lake and cooked lunch on a portable gas burner they carried with them. They didn't bother us and left an hour later. I was writing at my desk when Pasang returned from a tour of the moraines.

'Father, you must come. Big tracks,' he announced.

Well, the lethargy was something to overcome. But Pasang was so enthusiastic that I knew I had better show interest. I put the spotting scope and tripod into my rucksack, changed boots and set off with him for the moraine above the *yersin.* One hundred and fifty metres, I kept telling myself, was not a big deal to please Pasang.

From the top of the tallest moraine, Pasang pointed southward and handed me the 10 x 40 Dialyt binoculars. I swept the mountains below Gokyo. I couldn't see a thing. He pointed again. I looked at the knob of a mountain 5 kilometres (3 miles) to the south and tracks appeared through the binoculars so large that I wondered how I had missed them. Even without the binoculars I could now see them.

'Maybe two days old,' Pasang said.

I set up the spotting scope. The tracks led to a pass between the knob and the north buttress of Macherma Peak. On the far side of the knob, according to the map, were situated the yak pastures of Pangka and the hamlet of Macherma. Half the pass was already in the afternoon shade. The knob itself was still in bright sunlight.

The tracks began in a gully, out of view, between the first and second of the Dudh Pokhri lakes, which we had passed nine days before. They rose towards the pass, but two-thirds of the way up they split into two sets. The left set climbed towards the 5,056 metre (16,610 ft) summit overlooking Ngozumba's terminal moraine, circling below the summit to the top of a rockface. The other set continued over the pass.

From a distance of 5 kilometres it was hard to be sure, but these were not the tracks of a four-legged animal. This left two possibilities. Pasang assured me that no Sherpa would go there at this time of year. So they had to be the tracks of an energetic trekking couple, or a pair of yetis.

Pasang's discovery put Macherma back on the map as a place of interest. After my 1986 visit there I had discarded it as being too frequented by tourists. It had become a stopping place on the trekking route from Namche to Gokyo.

Our problem now became logistical. Although only 5 kilometres away, two days were needed to reach and investigate the tracks. But until Gyalzen's return we were anchored to our Donag camp.

Back at the Rock House the ermine had taken the cheese. The malt biscuits didn't interest him. At 3 p.m. a large black bird flew overhead, honking its way southward like a deep-throated crow. 'Gorak,' Pasang informed me. He explained that its home was the north side of the Himalayas, on the Tibetan plateau. 'Very rare,' he added.

At breakfast next morning the gorak returned north, wheeling over Coconut Beach, taking in the view of Rock House before continuing towards the *yersin.* His cry was different, like someone playing a Jewish harp. Not a cloud was in sight. Sweeping the horizon with the spotting scope to see if anything had changed overnight I noticed deep tracks coming off the shoulder of Gokyo Peak, just over a kilometre to the south. They had not been there the day before and I could have sworn that, like the tracks on the 5,000 metre Knob, they had not been made by a four-footed animal. I decided to investigate.

The spur was directly above the point in the ablation valley where on my way up from Gokyo I had found plumage. As I skied closer, I could see that the tracks came from a space between the crags on the side of the shoulder against which no snow had drifted, forming a perfect lair. I could now decipher what had happened. The tracks had been made by a cat bounding down from the lair to snatch a mouse-hare or other rodent whose imprints were now also visible. All four of the cat's paws had made each hole in the snow, so that they looked like the imprint of a two-footed creature and the stride was extraordinarily long because the cat, a full-grown snow leopard, was extending each bound to strike before its prey had time to react. The cat had come down twice and each time strolled back to the lair slowly, leaving normal paw marks. Could this be an explanation for other 'yeti' footprints? Perhaps some, I decided, but not those that continued across a barren landscape for several kilometres.

I took off my skis and lowered myself into the lair, hoping the animal was not there. It wasn't. There were a few tufts of hair, no bones and a beautiful view north and eastward. I wondered for how many days it had been watching us from its lair. Could it have been this cat that had displaced the rocks in the staircase next to Rock House? That certainly could not be excluded, though I thought it unlikely. Could this cat have made the imprints on the monticule or around the south ridge rock? No, those were too big, even allowing for sun-stretch. But it could have been the depositor of the scat that Gerard had found on our first day at Rock House.

Before returning to Coconut Beach, I climbed higher on the ridge and could have reached Gokyo Peak's north summit, but stopped after a few hundred metres as there was no point. Skiing back to Siphon Bay, I noticed that while I had been on the shoulder the cat had crossed my uphill tracks, leaving fresh paw marks. The sly creature was even then observing me. I searched the base of the rocks to see where it might be lurking but could detect no sign of it.

When I told Pasang, he took it very seriously. 'If snow cat, then also yeti,' he remarked. Gyalzen had told me the same thing when crossing Tshola Pass in April 1986. On the east side of the pass we had found tracks of a snow leopard that had led to a nest where grey fur clung to the compressed juniper bushes. 'Where snow leopard, also yeti,' he had said.[2]

After lunch I added a pancake to the rice on the rock not more than six paces from the boulder that served as my writing desk. I was hoping that the ermine would return but instead two yellow-billed choughs swooped in. They had already discovered the cheese and biscuits on the north side of Rock House, but they really liked pancake. They were quite unconcerned by my presence, but suddenly they fled, leaving an unfinished meal. This was unusual as they had voracious appetites. I waited to see that would happen. A minute later two goraks glided over Rock House and circled to land rather elegantly a few metres from me. They were huge, the size of thickset turkeys, with the wingspan of a Spitfire. I had never before seen anything that large quite as agile in flight. Their beaks were hooked, like parrots, but as black as their feathers. Their necks were short and they looked like pall-bearers on wings.

They made short shrift of the bait and moved down to the kitchen where they had a ball. They hopped onto the roof and poked their heads through the chimney vent, then hopped down and made their way around to the door. Every time one started to enter I whistled and it came hopping back out again. They turned the garbage tip upside down. Pasang was on North Slope gathering juniper branches so they inspected his tent as well. At times they would stop and coo over each other, the larger one plucking at the feathers of its mate or stroking the other's neck. It was touching: they really

[2]Researching the connection between the snow leopard and the yeti, Prof. Porshnev noted that the snow leopard, unlike the ordinary leopard, never attacked man. This instinct, he speculated, may have developed in the snow leopard because of a natural relationship between the large feline and the wildman-scavenger. This implied that the animal showed no aggressive tendencies toward man because for millennia the two had lived in close proximity, one depending on the other for a part of its food, perhaps even working together to flush out mutual prey (see page 16).

loved each other. They stayed for a good twenty minutes and, once they had picked clean everything they could get hold of, they rose up on their wings and flew out over the lake. As soon as they had gone, the choughs returned.

We had an early supper and I was back in my tent preparing, without much enthusiasm, for my evening patrol — I was sure it was giving me pneumonia — when about 6 p.m. I heard Pasang Tshering's voice calling from the *yersin.* I put on my headlamp and went to guide him and his father to the lake. They had come from Pangpoche in nine and a half hours.

We lit the kitchen fire and called Uncle Pasang to join us. The raksi had come. We brought each other up to date and while we talked a mouse scooted along the base of the wall. Pasang Tshering was beaming. His resilience was amazing. Next day, while Gyalzen recuperated, Pasang and I climbed Ngozumba Kalapattar (5,553 metres/18,242 ft) from where we could observe the passes east and west into the neighbouring valleys. Above Ngozumba Tsho in every direction was a white desert with no sign of life other than the choughs that had followed us. This was not yeti country.

Returning to camp, I again had pain in my right lung. But it was Gyalzen's health that most worried me. He said he felt 'very strong,' but his cough remained and he admitted to having dizzy spells. He looked frail and I wondered if the search for the yeti was having a psychological effect on him. Gyalzen, too, had thought of the yeti's curse — according to Sherpa lore, anyone who disturbs the yeti risks serious consequences, even death — for he announced that he had ten monks coming from Tengboche to pray for him. 'A big *pujah*,' he said. 'Two-day *pujah-bu*.' Big Pujah means Big Party in Sherpa country, with most of the village invited, after the exorcism, for the final day of prayers, drinking and dancing. Gyalzen didn't want me there for the exorcism, but asked that I and Pasang turn up around noontime on Saturday, 5 December. Next morning, Tuesday, he and Pasang Tshering left for Pangpoche. Before departing, Gyalzen said that on the following day he would send two yaks and a driver from Nah to carry our gear back down the valley.

On Wednesday, 2 December, our twelfth day at Donag, we dismantled the camp. The lake had dropped 6 centimetres (2.5 inches) during our stay and now was two-thirds frozen over. Lama Sanga Dorje's cave came to mind as I packed my rucksack and I thought about the monk's place in Sherpa society. He taught them that what they see *is* the reality, although it may be only a shadow. Looking at the phenomenon from a slightly different angle, in a culture that may have no more touch with reality than a steadfast belief in a hermit who could fly over Everest, hang his coat upon a sunbeam and from whose shorn snippets of hair grew a forest of juniper trees, who needs currency for commerce, TV entertainment, or food stamps to cash in on supermarket abundance?

The BBC had kept me in touch with the reality of the Occident while we had been camped at Donag. That reality said that the level of the dollar, like the level of the lake, had dropped. In fact it had dropped to its lowest level in more than forty years against other major currencies, depressing the Dow Jones Industrial Average once again to the mid-1700s. The economies of the West were in turmoil, while in Sherpaland there was enough potatoes, tsampa and sonam for everyone.

At 11 a.m. three porters — the yaks had already left Nah for winter pastures further south — arrived to help us descend the eight difficult kilometres (5 miles) to the far side of the 5,000 metre Knob. It was going to be a heavy carry. I added the skis to my rucksack and after a midday snack set out for Nah and, on my way by, a quick look at the footsteps on the northern face of the Knob.

My initial feeling on leaving Rock House was that our time at Donag had been a washout. If only one trekker had visited our camp it would have been too many, but during our twelve days there we had seen eight in close proximity so we were far from being isolated. This had kept us tied to our camp, not wanting to risk having equipment pilfered. And yet in spite of the lack of isolation there had been an animal presence, some of it explainable, like the snow leopard working a territory to the south. But other incidents were less so. The pair of midnight tsonkers, for example. Were they related to the footprints on the monticule or the North Slope saddle? I tended to think so, but this only deepened the mystery. Yetis are said to chatter, scream and whistle, but I knew of no large animal that tsonked. Curiously, goraks tsonked, though nothing suggested that they might be night flyers.

Were the midnight tsonkers linked to that other nocturnal visitor, the Rock House scamper? This I also thought likely, but again there was nothing to prove it. In any event, *something* had been behind Rock House that one night, perhaps even several nights; that much was undeniable.

The other undeniable fact was that Donag, with its jumble of moraines and monticules, glacial rubble, North Slope and back valley, not to mention Ngozumba's stony road, was large enough to hide a regiment of yetis without our ever realizing their presence. To be sure, our being camped by the lake would have been an inconvenience for them, restricting their movements. But by being careful, which they were, there was no need for them to leave fresh tracks, unless of course there was a fresh snowfall. When a change of wind and temperature occurred on 28 November, Pasang and I had expected a snowfall, though one never materialized.

Once Pasang showed me the tracks on the 5,000 metre Knob I became convinced that, anticipating a winding down of the trekking season from December to February, our quarry — a yeti couple — had departed down-valley. This conviction was reinforced by an absence of activity around Donag since the nocturnal visitor's scampette. As we had seen the Knob tracks on 28 November and Pasang estimated they were then two days old, everything coincided. After visiting us during the night of 25 November, nocturnal visitor and its mate had left Donag for an area where winter food was more abundant.

Chapter 11

The 5,000 Metre Knob

Walking down from Donag I had ample time to sift through the facts once again and ensure that nothing had been overlooked in that particular section of Lama Sanga Dorje's cave. I was now convinced that Monticule Man, Pond Creature and Nocturnal Scamper were the same animal and that shortly after its 25 November night-time visit to Rock House it had departed Donag with a mate to cross the 5,000 metre Knob. Below Gokyo, I could clearly see their footprints rising from the base of the Knob. I shed my pack by the stream connecting the first and second Dudh Pokhri lakes and focused the spotting scope on the vertical rockface I had seen from the moraine at Donag, looking for the chimney above which one of the animals had stopped. With the scope I could make out a cave that was well inside the chimney. So, I told myself, it was this cave that the animal had been trying to reach. It was probably looking for shelter after a nocturnal march, a place to hide before sunrise.

I crossed the stream and picked up the tracks where they started the climb to the pass at the right of the Knob's summit. The tracks were impressive. Though there was no way of knowing, I estimated they were a week old, which fitted the time frame perfectly. They had been dusted with spindrift and softened by the sun so that they looked as if made by a pair of stockinged feet.

I walked in the footsteps for 30 metres. Their stride was longer than the tracks around Rock House. The imprint measured about 35 to 40 centimetres — my boot fitted easily into them. But of course two animals had been in these tracks as they climbed one after another in Indian-file through deep snow. I quickly discarded the possibility that the imprints might have been made by a group of trekkers heading over the pass. No marks of ski poles existed on either side of the tracks, and there were none of the discarded Kleenex, plastic pill separators or toilet paper that inevitably accompanied trekkers.

Next morning Pasang had no enthusiasm for climbing to the top of the Knob with me. As he had spent the evening drinking raksi with a yak herder from Phortse, I assumed they had discussed the tracks and who might have made them. Over breakfast he asked, sheepishly, if I needed him.

'If you'd like to come, I'd like to have you with me. But if you prefer to stay, that's okay too,' I told him.

He mumbled something I wasn't able to catch and I asked him to repeat it. He wouldn't, but said instead that many people went over that pass, all the time. This contradicted what he had told me four days before.

'What sort of people?' I asked.

'Tourist people. Many groups.'

I considered that pure hogwash. Pasang was scared of encountering the yeti.

'Tomorrow we explore Macherma valley?' he suggested.

'Maybe,' I said, and set off for the Knob.

The countryside under Ngozumba's terminal moraine is by no means lush, but immediately it appears more hospitable than the tundra around Donag. On its southern side much of the Knob is covered with grass on which yaks graze in summer and from its base stretches a gently sloping shelf that leads to the more expansive pastures of Pangka. The route I chose up the mountain was a 50-degree corridor between two rock outcroppings. A cloud of choughs remained above me most of the way until I reached the summit snow at midday.

The day was warm with hardly any wind and the sky was a deep blue that silhouetted the snow-capped peaks on all sides of me. At my feet Ngozumba's rocky road began its climb to the northern horizon, ending against the ramparts of Cho Oyu at an altitude higher than the mountain on which I was standing. From a summital cairn I descended to the Knob's northside rockface and easily found the tracks that headed to the chimney where I now knew a cave was hidden. The animal had gone into the top of the chimney, which was laden with snow, obviously wanting to reach the cave, but I could imagine it became concerned about starting an avalanche. I knew I was.

Had I a rope and Pasang with me I would have belayed down to the cave. I followed the creature's tracks, much smaller when made by a single animal, into the four-metre-wide neck of the chimney and saw, two metres from me, where it had stopped, dropped to all fours, turned and climbed back out, traversing under the summit to rejoin its mate. I did the same traverse and got to the point where both of them had crossed the pass. A broad corridor continued down the south side, exposed to the sun and therefore without a flake of snow on it. No more tracks.

Where the animals would have come out of the corridor was a band of rock debris deposited by one of Macherma Peak's hanging glaciers before it withdrew. They had to cross this band to reach, directly above Pangka, a snowy slope that led to the rounded shoulder of Macherma valley. On the far side of the rock debris the tracks resumed in the snow, heading for Macherma, or rather for another cave that was at the base of a rock buttress above the hamlet of Macherma.

I descended into the corridor. It was a mixture of sand and scree and the footing was not good. I zig-zagged down but found no tracks, no signs of Vibram-soled trekkers, no Kleenex, no toilet paper, no plastic pill separators, nor any bubble-gum wrappers. It was late and rather than cross the debris to follow the tracks toward Macherma, another two kilometres away, I continued down the rock gully towards the pastures above Nah. This proved to be not such a bad decision.

I noted in the gully that heaps of boulders formed cave-like shelters, some even inviting. I went into the largest of them. It was big enough to accommodate several people my size. On the ground there were millions of little brown pellets — bat shit — a sampling of which I scooped into a film capsule. But, more interestingly, somebody had laid a bed of juniper branches in one corner of the cave. This 'somebody' obviously could not have been a bear or snow leopard. It had to be 'somebody' who was smart with his hands — perhaps a yak herder, perhaps . . . a yeti, perhaps . . . Monticule Man.

Continuing downhill I saw a distant figure walking towards me. It was Pascal Marlinge, the Frenchman who worked with Yeti Mountaineering and Trekking as a guide. He had arrived at Pangpoche from Kathmandu with 12 kilograms of sugar and 10,000 rupees from my account with the agency. He had spent a day at Pangpoche and thought that Gyalzen had aged ten years in two months. He said Pangpoche was a busy place as preparations were in progress not only for Gyalzen's pujah but for a wedding, the groom and his family coming from Phortse with a barrel of chang, as tradition would have it. He said Gyalzen appeared weak and was unable to get out of bed in the morning. He also said Gyalzen was talking about changing the programme, as a herder with pastures at Chhukhung had reported sighting fresh 'yeti' tracks near a summer *yersin* called Shanjo, nine kilometres up the Imja valley from Pangpoche. When we arrived back at the camp, I questioned Pasang Tshering, who had guided Pascal to Nah, about his father's health. Papoo was very well, he said.

The BBC announced that evening that seven European central banks had lowered their interest rates to bolster the dollar. The US Treasury, by allowing the dollar to slide in world markets, had forced America's trading partners to move away from conservative monetary policies. Stimulating a global consumer economy with easier credit while rolling over uncollectable international debts may have been Washington's logic, but in Khumbu this was not a logic that the delicate environment could endure.

Tin cans and plastic bottles already littered the streams

and pastures. Soon the merchants of Namche would be importing microwave ovens and bigger, better VCRs with more blue video cassettes to attract the tourists. And should ever the yeti be found, it risked becoming a hotter tourist attraction than the sunrise on Mount Everest, maybe even revitalizing the fortunes of Everest View Hotel, removing it from limbo and putting it back in the promotional literature of international tour operators. Somehow this didn't seem right.

That morning at breakfast I had a used headlamp battery in my hand and didn't know what to do with it. The battery, a 4.5 volt alkaline long-life job, had died on me while I attempted to bring my journal up to date during the night.

'What do I do with the battery?' I asked Pasang, not wanting to carry it back to Pangpoche and hoping that maybe Nah had a magic incinerator. Such a state of mind is known as Himal Deliria.

'Throw it outside,' Pascal replied ingenuously.

Garbage in, garbage out is the basic mountaineer's rule. But here 'garbage out' meant carting it back to Kathmandu and even there they didn't know what to do with it.

Chapter 12

Gyalzen's Pujah

During the last ice age Ngozumba Glacier almost certainly progressed as far south as Phortse and beyond, just as the region's two other major glaciers, Khumbu and Nangpa, also descended much lower in their valleys, flowing south from the central spine of the Himalayan range. On either side of the giant Ngozumba, lateral glaciers flowed into it: at each of the five Dudh Pokhri lakes, for example, and at Macherma as well, coming from a high valley wedged between Macherma Peak and Kyajo Ri. But these lateral glaciers had receded much faster, leaving moraines that have been softened in form and texture by eons of erosion, like Donag's North Slope and the shoulders of Macherma valley. In fact these former moraines have been sufficiently eroded to leave a very rudimentary deposit of top soil, enough to support scrub and hardy grass.

Macherma valley rises in two steps from the floor of the Dudh Kosi trough. The first step is broad and alluvial, containing the hamlet and best pastures. Its backwall is an abrupt cliff on top of which begins a second, higher valley, still glacial, which curls behind Macherma Peak, hiding it from the view of anyone standing on the first step. Three glaciers remain in the high valley and the melt-off from them becomes a stream that runs from the base of the cliff through the lower valley to empty into the Dudh Kosi. Macherma itself is a collection of not more than a dozen widely dispersed yak herders' houses, and the whole valley, rocky though it is, is used as a grazing range for free-roaming herds. The area has a rich history of yeti activity, but the most notorious incident occurred there in 1974, involving a 19-year-old Sherpani. Her name is Lhaupa Dolma and she comes from the village of Khumjung.

On 12 July 1974 Lhaupa Dolma was tending her parents' mixed herd of yaks and zoms. A zom is the crossbreed between small Tibetan cattle (*Bosaunus typicus*) and a nak (female yak). A zom can live at lower altitudes and gives more milk than a nak. The male of the crossbreed is called a zopiak and, unlike the zom, it is not fertile.

It was mid-monsoon and raining. Under her cape, Lhaupa Dolma was thinking of her brother, a climbing sirdar who was due back any day from Everest Base Camp, when she heard a whistle. She immediately thought it was her brother arriving from Khumjung to see her and she sat by the stream to wait. Minutes later she was grabbed from behind and flung into the stream.

When she regained her senses Lhaupa Dolma saw a chuti, the large type of yeti that has a well-developed taste for yak meat, grab one of her zoms by the horns and kill it by snapping its neck. The chuti then ripped open the zom's stomach and helped itself to a handful of entrails. Lhaupa Dolma did not move, but lay in the glacial water as if dead and watched as a baby yak came along to see what was happening, followed by its mother. The chuti threw the calf six metres, killing it, and broke off the legs of the nak to suck the marrow from its bones. It tried to hide one of the carcasses under a rock overhang and then, having gorged itself, sauntered away into the rocky wastes, climbing at a fast pace up the southern base of Macherma Peak, sometimes on all fours, sometimes erect.

After the chuti was out of sight, Lhaupa Dolma ran back to Macherma and collapsed into the arms of another yak herder, trembling so much that she was unable to speak. Finally she regained enough composure to tell the herder that she had been attacked by a two-metre tall chuti with shaggy brown hair that parted at the waist. The hair fell downward on the lower half of the body, she said, while above the waist it grew upwards. She also remembered that it had a longish head and walked pigeon-toed when on its hind legs. The herder went to see the carcasses for himself and then took Lhaupa Dolma down to Khumjung, where her mother Dhar Dolma had the presence of mind to send for the village lama. The Khumjung lama performed 'a powerful pujah' to exorcise the spell of the yeti.

The police in Namche were informed of the attack and opened an investigation. A police officer went to Macherma, saw the dead animals and was shown the tracks that the yeti had left in the sand by the stream and further up in the mud of the pasture. He drew a sketch of the tracks, photographed them with his box Brownie and filed a report on 15 July 1974. The pictures were disappointing but the report was vivid. Yetiologist René de Milleville told me he had read the police report and interviewed both the officer and Lhaupa Dolma shortly after the incident. The police report as well as the interviews, he said, substantially confirmed Gyalzen's version of events.

The story, as told to me by Gyalzen, seems to contain some illogical fancy and the animal behaviour he describes certainly would appear to be an inefficient use of energy for a large predator, but in general the story jibes with the other sources I consulted.

Everyone I spoke with about the incident was convinced that the powerful exorcism performed by the Khumjung lama saved Lhaupa Dolma from succumbing to the curse of the yeti. She had a nervous breakdown that lasted three months but recovered. She still lives in Khumjung, now married and the mother of three children.

Pascal Marlinge and I climbed from Nah to the shoulder above Macherma where the yeti tracks from the Knob crossed to a network of caves at the base of Macherma Peak's south-east buttress. From there we could see other tracks in the valley below, some traversing diagonally, others heading up to or descending from the cliffs in the north-west corner of the valley, obscured from our view by other buttresses descending from Macherma's fluted summit. I knew by then enough about yak tracks — yaks drag their feet, leaving troughs in the snow between each imprint — to be able to distinguish which had been made by yaks and which had not. Those that had not were mystifying, but rather than descending to examine them we decided to explore Macherma Peak's southern ramparts and particularly the caves at the base of the south-east buttress.

The tracks from the Knob were at this point non-identifiable depressions in the snow. But clearly they indicated a pattern of two-legged animals coming from the Knob. The caves were connected by corridors and chimneys. One had a frozen spring with crystal-clear ice, but no signs of past or present tenants, while on a rocky roost in front of another I found a pile of bones, pieces of jaw, skull plating and tibia which I placed in my rucksack to show Pasang.

We continued along the south flank, hoping for a view into the upper valley, until discovering a hidden amphitheatre at nearly 4,900 metres (16,100 ft), with a Mediterranean micro-climate, caused by the sun reflecting off its smooth granite slabs. In the faults between the sky-scraper slabs were caves with vaulted ceilings like gothic cathedrals, just out of our reach without climbing gear. While making my way to a higher platform on the east side of the amphitheatre, I glanced over my shoulder to see a griffon vulture with a four metre (12 ft) wingspan launch itself off the ledge next to me and glide out of sight behind a spur. On another nearby shelf Pascal found fragments of jaw and thigh bones similar to the ones I had found earlier. They were friable and almost ready to disintegrate into dust, though when later I showed them to Pasang he said they belonged to a baby serow, or mountain goat, that he estimated had died two years before.

We sat for an hour in the hidden cirque, looking down on Macherma and its valley. Hundreds of metres below us, five yaks, tiny black spots, were browsing where the sparse pastures were bare of the October snow. I could imagine the chuti that had attacked Lhaupa Dolma escaping to just such a refuge, losing itself in the dark rock faces where nobody would dare pursue it.

We arrived back at Gyalzen's house at mid-afternoon on the following day to find the pujah in full swing. Gyalzen had erected two tents in the potato patch for Pascal and me. The trunks of two trees and a row of firewood were stacked in front of the house while on the back porch tsampa was cooking in a large copper cauldron over an open fire by the platform that served as the toilet. About forty people were crowded into the long room and ten monks were reciting prayers in the lhang. Seated cross-legged on the banquette by the hearth, in the place of honour, was an old man with long white hair. This was Karma, Kanche's father, whom I had met the year before at Bhandar where Kanche's youngest sister, Kaaputi, whose husband was killed on the 1985 American Langtang Lerung expedition, lived with her five children. Next to Karma was another older man, with grey hair tied into tresses by ribbons, dressed in a *chuba*, the traditional Tibetan coat, and a Tibetan fur-lined hat. This was Dawa Namge, Gyalzen's father. I, in my dusty coveralls, was guided to a place on the banquette between Dawa Namge, who thumped me on the back, and Ang Dorje, the nextdoor neighbour.

The men, wearing hats, all sat in a line upon the banquette according to strict etiquette. The women and any children they were nursing sat on mats spread along the floor opposite us. Kanche, aided by Yangtse and two Sherpani friends from the upper village served chang, raksi, butter tea and spicy tsampa à-go-go. Music was provided by the chanting of the monks in the lhang, and Gyalzen, wearing a chuba and tam-o'shanter, supervised the proceedings from the raised dais of his bed.

Kanche came over and filled my glass with chang. '*Shey*, Father,' she ordered. She would not leave till I had drunk two glasses. '*Shey, shey*,' she kept insisting, and if I showed signs of flagging she would push the glass towards my lips.

With Ang Dorje it was the same procedure. He tried to

refuse. This was only a ploy. In fact, it is good Sherpa manners to refuse food or drink up to three times so that the hosts can show how insistent they are that you partake of their hospitality. Kanche filled his glass again. '*Shey, shey,*' she repeated, and when he didn't drink quickly enough she put down her chang pitcher and raised the glass to his mouth, forcing him to gulp down the chang or spill it over the front of his fine costume.

Kanche was known as a generous but insistent hostess, though the procedure was more or less identical at other pujahs I attended. A pujah, big or small, centres upon petitioning the gods with appropriate offerings to perform some service on your behalf, at the same time making offerings and threats to the demons. While this is going on, much chang and raksi is consumed, the mortals drinking on behalf of the gods and demons. Those attending the pujah must eat and drink in great quantity if the gods are to feel inclined to assist in booting out the demons who by then are hopefully good and drunk.

Pujah is a major event in Sherpa village life. The biggest pujahs are held on occasions of births, marriages, illnesses or death. The variety of malicious, aggressive, violent and otherwise downright unpleasent beings who are invoked at these gatherings can be overwhelming and, to any outsider, at all times confusing. Even the highest gods can become nasty if they are not sufficiently bribed with offerings of food and prayers. Consequently, pujahs are concerned with combatting the evil moods or potentially angry dispositions of the Sherpa gods and spirits.

After more butter tea, chang and tsampa, each guest was given a plate containing a mountain of ritual food, the centrepiece being a pyramid of rice, known as a *tso*, its top dyed crimson. The tso is surrounded by cookies, candies, suntallas, peanuts and Sherpa doughnuts. Whatever the guests were unable to consume they dumped into a plastic bag brought with them for the occasion, so that it could be taken home and eaten later.

The plastic bags all seemed to carry the trademark of some

foreign sporting goods manufacturer, airport duty free shop, pharmaceutical combine, film manufacturer or supermarket chain. Unless ripped or otherwise rendered useless, imported plastic bags, being superior in quality to Napalese-made ones, are one of the few items unlikely to be found on any Khumba refuse heap. They are highly valued for carrying pujah leftovers, but also offerings of grain and rice, for burying winter potatoes in outdoor holes to preserve them from frost, for transporting market purchases of loose sugar or flour – their uses are almost unlimited. Equally prized but harder to come by are the plastic expedition drums, excellent for making pujah-quantities of chang or pickled Sherpa coleslaw. Kanche had already taken over three of my family of drums for such purposes and had her eye on at least two more.

After dark Pasang Tshering came running up the stairs into the long room to announce that Uncle Pasang had arrived and was unloading the two yaks we had after all been able to hire in Nah. But Pasang would not come upstairs for tsampa and chang, sending word that he had to go home to see his wife, Nima Yanchin. I asked Gyalzen whether something was wrong. Though I didn't receive an answer I already had an inkling that Gyalzen's sister and Kanche did not see eye to eye.

About 9 p.m. the lamas took a break and came back to fresh butter tea, chang and raksi. They appeared to be having a good time – their second day of reading prayers, tossing rice and burning incense. Once the prayers finished, the dancing began: the elders forming the centre of a long line, with all the women on the left, the men to the right. The dance, accompanied by a chant, was a slow, three-step shuffle, repeated backwards, forwards and sideways. The bagman among the lamas – for the lamas must be paid for their services – was a tall, robust fellow. He had heavy eyes from all the chang and prayers. With a smaller monk, he joined the dancing, but within moments was falling asleep on his feet and had to be supported until finally he passed out.

Until that evening I had considered Sherpa alcohol,

whether chang or raksi, to be mild, almost incapable of making anyone drunk. I was mistaken. Gyalzen, though himself unable to walk a straight line, led me to my tent towards midnight and I was told the dancing continued until two or three in the morning.

When I awoke next morning I had one of the most monumental headaches of my career. The bagman had spent the night in the lhang. When he saw me appear for milk-tea, he roared with laughter. 'Not feeling well?' he asked, taking me by the wrist as if to feel my pulse. 'You need a glass of chang,' he said.

I recalled having read a Sherpa proverb which translates: 'The rich get more money and the drunks get more chang.' I repeated it as Gyalzen handed me my glass. Everyone laughed.

That afternoon Pascal and I accompanied Gyalzen to the gompa in the upper village to return the two large trumpets borrowed for the pujah. The gompa was supposedly built around 1860, replacing the earlier one founded by Lama Sanga Dorje. The site was especially holy because of the many miracles Sanga Dorje is said to have performed there. Lama Sange Dorje's ascension to celestial bodhisatship is celebrated each year during the summer festival of *dumji.* Sherpas come from all over Solu-Khumbu for a week-long orgy of dancing and feasting during which the gompa's prized yeti scalp is brought out and paraded about the village.

Outside the gompa's main prayer room are paintings of the protective demons as well as of the gods of Chomolungma, Khumbilia and Taboche. The only painting of Sanga Dorje was inside the prayer room to the left of the alter, by a window. The yeti scalp and a skeletal hand are kept in another prayer room upstairs. For a 10-rupee contribution to the gompa's upkeep, anyone is allowed to view them. The scalp looks as if it has been kicked across the gompa courtyard a few times in drunken dumji scuffles. It has only a few reddish hairs left on it. Gyalzen blames this on the tourists, whom he said pluck out the hairs as souvenirs.

On Monday, 7 December, with Pascal Marlinge we crossed the Imja, intending to spend a couple of days at Mingbo, the pastures four kilometres east of and 660 metres higher than Pangpoche, near the base of Ama Dablam. Gyalzen was now convinced he was cured — the power of suggestive medicine. He walked in a spritely fashion and once again with confidence. We soon outclimbed the clouds and the going was much easier than three weeks earlier. The day was warm, but I knew that as soon as the sun dipped behind the mountains it would become bitterly cold, so I started gathering juniper branches and dried yak dung for the rock-shelter kitchen while waiting for Gyalzen and Pasang to arrive with the yaks. At 4,560 metres (14,960 ft), Mingbo is lower than Donag but no less cold. We wanted to see if there were any new tracks since our last visit. Mingbo's major shortcoming as a winter camping place was the fact that the sun only reached it after 9 a.m. Without the sun it was an icebox. Also the nearest unfrozen spring was twenty minutes away.

At night the moon turned the north faces of the mountains phosphorescent Around the fire in our starlit rock-camp kitchen, Pascal explained why he had rejected the 'society of waste' and turned to Buddhism. In many ways he was the opposite of Gerard. He had been a pharmaceutical salesman in France before his marriage broke up and had hated being part of the rat race. He left behind a seven-year-old daughter whom he was financially unable to support. Himself the product of a broken marriage, he was brought up by his grandparents. He earned $1,200 a season from guiding French trekking groups. He had learned Nepali and lived with a family in a distant village, which cost him next to nothing, for he was a vegetarian and ate only dal bhat.

Over the next three days we explored the fractured wasteland around the base of Ama Dablam, an immense shaft of paleozoic rock that about 300 million years ago began its upward thrust from the Sea of Tethys that then covered Khumbu and the Tibetan plateau. It was incredible to think of the gigantic forces that uplifted this shaft to such towering heights. We climbed towards its south-east shoulder for a view of Nare Glacier's screefield and the icy desert to the east that stretched to Mingbo La. Even in this barren wilderness at 5,250 metres (17,225 ft) we found yak dung, indicating the existence of food for summer grazers. Tracks were visible on both sides of the shoulders, but so old that they offered us nothing.

On 10 December we left Mingbo in two parties: Gyalzen, Kanche and Pascal with the two yaks, Marsan and Tsao, via the normal route; myself and Pasang descending to the Nare Drangka, crossing it and climbing a 200 metre snow-filled corridor behind Rawldurche. The corridor was on an incline of a good 60 degrees towards the top and ice axes were needed, but we were rewarded by finding old tracks on the upper saddle. Some quite certainly had been made by tahr, as we found tahr droppings, but Pasang thought others had been left by one or more yetis.

Our strategy, conceived by the campfire the night before, was to send one party — Pasang and myself — down the valley under Rawldurche's main summit, driving any animals ahead of us, while a blocking party — Pascal — would take up position at the bottom of the valley, to observe anything that might exit. Giving what we estimated was enough time for Pascal to get into position, we started down. We had to hop or jump from boulder to boulder, being careful not to start a rockslide. Occasionally the boulders were interrupted by a line of scree which was no more amusing to cross.

The main Rawldurche ridge was now to our left. We were following the base of a northern spur back down to the Imja Khola, almost opposite Pangpoche but in fact a bit to the north. It was windless and the sun was hot. One cave I visited was large enough to stand up in. It contained a fair amount of tahr droppings. Pasang would not come into the caves but wandered further down the valley to wait. We knew at one point we would have to descend a cliff and this worried us because, as it was not yak country, Pasang did not know whether a passage existed. Coming out from one rock overhang, I caught a fleeting glimpse of a reddish-brown shape moving down the valley 200 metres below us.

'Yeti?' we asked each other simultaneously. But another glance before it disappeared between the rocks told us, we thought, that it was a serow, or wild mountain goat. Either it was going to climb through a breach in the rockface to the ridge above or continue down towards the valley entrance where Pascal would see it.

When we came to the clifftop we debated which was the safest descent. We heard two yelps and down on the north moraine was Pascal, sitting on a flat rock in the sun. We traversed to the south side where the cliff was more eroded, but found a frozen waterfall too treacherous to afford a passage. A route around it posed another set of problems, but these we overcame at some risk, getting to the bottom and crossing back again toward Pascal.

'Did you see the man with the mouflon?' he asked.

We were gazing across the 500-metre-wide bed of the former glacier that lay between two of Rawldurche's arms. It was an unfriendly place where not even yaks ventured as there was nothing there for them to eat.

'What man?' I replied.

'First this mouflon came down the valley, all furry with a nodding head. It was followed by a gangly man. I assumed it was Pasang. He had no rucksack, was walking fast and I had the impression that he was wearing the same brown-coloured clothing. I just glanced quickly at him with the binoculars. He was coming down the centre ridge, between the large rock and the pile of magma,' Pascal said, pointing to the remains of a median moraine in the middle of the valley. 'Immediately I looked for you. I thought you might have had an accident as Pasang had no rucksack. Then I saw you on top of the rockband and Pasang was behind you. I looked again for the gangly man. I saw the mouflon heading for the pasture at the bottom of the valley. But I couldn't find the man again.'

Pascal had brought a thermos of tea with him. We sat in the sun and discussed what we should do. We decided to converge on the arched middle moraine — a sort of dorsal fin formed from the debris when this pocket-sized glacier melted at the end of the last ice age, about 15,000 years ago — at three different points between 'the large rock and pile of magma'. If Pascal's gangly man was still there we would then be astride his route. If gangly man did not show himself we would scour the south side of the dorsal fin, sheltered from the sun by the shadow of the mountain, where there was still snow, to see if we could find his tracks.

We went as planned but found no tracks, much to Pasang's relief. A few rock heaps with spaces under them might have offered cover to even quite a large animal. Pascal crawled into two of them, but they were empty. So we continued down to the *yersin* at the bottom of the valley.

It was then 4 p.m. and the late afternoon was unaccustomedly warm. As clouds were drifting across the mountains from the west, Pasang predicted a change of weather. We returned to Pangpoche and decided to leave for Shanjo in the morning. That night the BBC reported that the US had a record October trade deficit of $17 billion, provoking new Wall Street jitters and a further decline in the value of the dollar. There is no word for *billion* in Sherpa-ka. Sherpas don't know how to count that high.

Chapter 13

Shanjo

I don't know what Gyalzen really thought about Pascal's story of the gangly man. I think we all felt uneasy about it, Pascal himself didn't mention it again.

Next day we crossed the Lobuche Chubung (a Tibetan word for stream) below Pheriche, six kilometres north of Pangpoche, and climbed along the right bank of the Imja to the village of Dingpoche at the foot of Pokalde Mountain. The Erwin Schneider map of Khumbu Himal shows the south-facing slope above Dingpoche to be partially wooded, based on the German cartographer's 1955 fieldwork. It was in a forest above this village, where today a gompa stands, that Sanga Dorje first hid from King Tshomtomba. But the slopes above Dingpoche are today totally bare, the trees having long since departed in smoke, up the chimneys of Sherpa houses.

Dingpoche was once Khumbu's most prosperous village, but after the forest disappeared the village's fortunes declined. It became all but abandoned in wintertime, to be inhabited again each spring. With the development of trekking, and the cash flow it provides, this has changed somewhat. In recent years two lodges have remained open more or less all winter to cater to the hardy tourists who come to admire the fortress-like black wall, with its gold-coloured seams, of Lhotse's south face.

When the glaciers retreated, they left a deposit of sandy loam at Dingpoche's doorstep and one result is the unanimous agreement among Sherpas that Dingpoche produces the best potatoes in Khumbu. As a Frenchman can tell the appellation, and sometimes the millésime, of the wine he is drinking, many Sherpas biting into a boiled potato can tell you, 'Ah, Dingpoche *alu*,' or 'Very good Khumjung *alu*.'

Clouds were being blown in from the west when we reached the village. It had been much milder overnight, and this was one of the warmest days yet. At the upper lodge, where we stopped for tea, the yard was filled with freshly cut logs. I asked the lodge owner from where he got such fine wood. From the forests opposite Pangpoche, he said.

'That should just about last you the winter,' I noted.

He agreed, pleased that I should care.

We left him and continued along the right bank of the Imja, heading east. Before Bibre, we left the trail and crossed the Imja's four fingers. We could see a long line of tracks descending in the snow towards the eight stone huts of Shanjo from the plateau around Ama Dablam's base. These were the tracks that a Sherpa herder supposedly had spotted two weeks before.

The Shanjo tracks were similar to those that crossed the 5,000-metre Knob above Macherma, but still I was unconvinced that they had been made by a yeti. Ama Dablam's north base camp was beside a frozen lake on the upper plateau, and it had been occupied by a Bulgarian expedition until the end of October. The map indicated that the line of tracks followed a trail leading from the plateau's yak pastures to Shanjo. My conclusion was that when the Bulgarian camp was evacuated, some people had come down by this route and these were their tracks.

Our destination was a rock shelter owned by Pasang's sister which he had promised would be comfortable. After clearing out the snow that had drifted under the door and infiltrated through the chinks in the dry stone wall, it became our kitchen. We unloaded the yaks and turned them loose.

'How can you be sure Marsan and Tsao won't wander back to Pangpoche?' I asked Gyalzen.

'Ah, because they know there is good grass at Shanjo and I

will feed them well,' he replied. By grass he meant hay, which he stored in a stone house he owned down by the river. The yaks headed in that direction and waited by the house until Gyalzen had time to feed them supper. They would get the same measure again in the morning.

As we pitched out tents in a corral beside the rock house, clouds began to descend around Nuptse and Lhotse. Even Island Peak, the white ship at the eastern end of the valley, became lost from view. The valley looked as if it once had been a battlefield for the seven major glaciers that converged at its upper limit. Piles of moraines, some higher than 5,000 metres, were lined up from Shanjo to Chhukhung like furrows in the sand.

We were treated to a furious sunset that turned the storm clouds over Taboche and Tshola peaks brilliant crimson, orange and purple. By 6 p.m. it was dark. Around 1 a.m. it started snowing. The cloud ceiling dropped below 6,000 metres. At 4 a.m. I had heard a rock fall from one of the dry walls on the south side of the tent and crawled out to investigate. I thought I heard footsteps and swept the riverside with Orion, but was unable to detect anything in the light-falling snow.

After fifty-two days, my prayers for fresh snow had finally been answered. Before dawn I was out looking for new tracks. The cloud ceiling had fallen below 5,000 metres and it seemed I could almost touch it. Later in the morning we walked up the rocky bed of what once had been Ama Dablam glacier. It was as if the gods had sprinkled boulders about like salt and pepper. They were piled haphazardly on top of one another with gaping holes in between and overhangs as large as walk-in closets to hide in. In front of one particularly spacious overhang with a sandy floor some smart animal — man or beast — had placed slabs of slate as a shield against the elements. This reminded me that the Sherpa word for yeti was *yeh-tey*. *Tey*, once again, is the general Tibetan word for animal and *yeh* means rocky places. In other words, a yeti was a tey that lived in rocky places. Here, then, was its playground.

For the next three days we noted that, except for a fox and a family of mouse hares, the animals that had left a profusion of tracks on the upper plateau during the fifty-two-day interval between snowfalls had now departed. No new tracks appeared.

On 16 December we also departed, leaving a stock of material hidden under the hay in Gyalzen's stone house. We arrived back in Pangpoche to find that Mingma Ramu had come down with chickenpox. Next day we said goodbye to Pascal who was returning to France for the holidays. That afternoon one of Gyalzen's neighbours appeared in the long room. He wanted to tell us a story.

In Sherpa society there is little visible form of government and yet everything seems to work with haphazard order. The only police in Khumbu are the Nepali police sent by the King to enforce the King's laws. Otherwise Sherpas have lived for centuries without police. They have strict taboos and any breach of them, such as intermarriage, thievery and murder (the latter all but unknown among Sherpas) is strictly dealt with, the offenders usually being banished. Though there are no lawyers, notaries, notarial deeds, land registry office, wills or codicils, Sherpas have enormous respect for private property. How they keep track of land titles I'll never know. But most people, like Gyalzen and Pasang, owned pieces and parcels here and there, such as at Taboche, Mingbo, Shanjo or Bibre.

Our visitor owned a rock camp under Konar Pass and another rock house at Deboche, close to Tengboche, on the left bank of the Imja Khola. For several years he had noticed that every January a yeti crossed Konar Pass from the Dudh Kosi valley, and spent a night at his rock camp above the Imja Khola, which was left vacant during the winter, visited only as snow conditions permitted. After spending one night and perhaps a day at the rock camp, he said, the yeti crossed the Imja south of Pangpoche to a wooded gully on the far side, where it rested again before traversing through the forest to the pastures above Deboche. He said he had never seen the animal, but always knew when it had visited the rock camp by the strong odour it left.

His story made good sense in terms of animal patterns and living habits. The absence of recent tracks in the higher spheres around Gokyo, Donag or Shanjo would indicate that the yeti left the high pastures after the first snowfall and migrated to forested areas where adequate food and shelter existed to survive the winter. We thanked him for this information and decided to visit the eastern approach to the pass next day.

That night a BBC news bulletin announced that Ivan Boesky, the New York arbitrageur, had been sentenced to a three-year prison term for insider trading. Boesky had already paid a $100 million fine. That represented one-sixth of Nepal's annual budget, or by another measure a sum that size could have provided the city of Kathmandu with sewage and garbage disposal systems, and clean drinking water, and brought electricity to much of the valley's outlying areas.[1]

When I tried to explain this to Gyalzen, who in his whole life had been as far north as Tingri, across the Nangpa La in Tibet, and as far south as Britnagar on the border with India, these were proportions — or disproportions — with which he was unable to grapple.

'How many yaks would that buy?' he asked.

'More than exist in all of Khumbu,' I replied.

'How many?' he insisted.

At $120 a yak I did a quick calculation. '833,000,' I told him.

He was used to counting in crore and lakhs when the numbers got above one hundred thousand, which wasn't very often.

'Are there that many people in the world?' he asked.

I assured him there were. Then he wanted to know what was insider trading. This I found more challenging. 'Imagine,' I began, 'that the police chief in Namche is informed by his radio link with Kathmandu (there are no telephones in Solu-

---

[1]Nepal's 1988 budget was 15.2 billion rupees.

Khumbu) that one thousand trekkers have left Jiri for Namche. With this information, which nobody else has, the police chief withdraws enough money from the bank to buy all the sugar on sale at the Saturday market, knowing that when the trekkers arrive they will need sugar and he can sell it to them at four or five times the price.

Gyalzen didn't see anything wrong with this. The official was merely exercising good business acumen, even though he was charged with regulating the Saturday market and insuring that no one was cheated. In fact Gyalzen thought the official would have been stupid to have missed the opportunity.

Next morning on our climb to Konar Pass I made an important discovery. By all accounts yetis have high-pitched whistles. But I found that they are not alone in possessing such talent. Tahrs also can whistle! We surprised a herd of them coming off a snow-covered ridge to drink and cavort by a stream under the pass. We crouched behind some rocks and watched as one steer bounded down the hillside in pursuit of a reluctant cow, trapped her by a large rock and they coupled. The herd by then had seen us. The junior steers whistled at their cows and calves and looked to the lead male for instructions on whether they should move away.

When I heard the first whistle I couldn't believe my ears. It was shrill and human-sounding. But visibly we were the only people on the mountainside. It must be a yeti, I told myself.

'Who whistled?' I asked Gyalzen.

'Jharral,' he said. And sure enough another buck let out a short, searing blast. The closest ones were only 20 metres from us. We observed them for three-quarters of an hour till they returned towards the ridge and we continued our climb into the clouds, that were forming around the top of the pass. We reached a window above the pass, could see no tracks, and began our descent. Back home we found that Mingma Ramu was better, but Tsheden had an earache. Gyalzen immediately took her in his arms and rocked her to sleep.

On my first visit to Gyalzen's home, Tsheden had been the baby at her mother's breast. We had arrived after nightfall and the front door was already locked. Candles were lit, Yangtse came down and opened the door while Kanche blew at the embers to induce the flames to return so that she could brew us tea. While Kunga Pemba sought his father's praise for having been a good boy — Pasang Tshering was then at school in Tengboche — and Yangtse bounced back and forth aiding her mother, Gyalzen only had eyes for Tsheden. He cooed at her, kissed her and carried her about.

Pasang Tshering was for Gyalzen the brightest boy in all of Khumbu. The family's spare cash went to pay his tuition at the monastery school where the little antelope boarded during the week with other novice thawas and returned home on weekends. The tuition was 300 rupees ($12) a month, which was a lot for a Sherpa family.

Tuition at the village school, on the other hand, was free. But the schoolmaster was Nepali. The government had forbidden the use of Sherpa-ka in Khumbu schools (monastic schools excepted) as part of its efforts to promote a national identity. This was not an issue. The Sherpas accepted it. Education in general was not very high on their list of priorities. Gyalzen's concern for Pasang Tshering's schooling was in this respect an exception. But for Kunga there would be no secondary schooling and it mattered not that the Pangpoche schoolmaster had only one dank classroom in which to teach all of the village's thirty or so children of primary school age. And while the Pangpoche school was all right for Kunga, for Yangtse, future housewife and mother, education was not important at all, even unnecessary.

Pasang Tshering had become the parents' prized pension plan, an investment in the future, though according to Sherpa custom the youngest son received the family house and in return had to care for his parents in their old age. By Kunga Pemba's attitude it was evident he felt equal in every respect to his elder brother and even at the age of nine or ten wanted the same opportunity.

I had gathered by then that the interval since my first visit had been an economic disaster for Gyalzen and Kanche. I was not informed of the reasons for this. I could only observe the consequences. Because he had been to the monastery school, Gyalzen was fluent in Tibetan and could read Sanskrit as well as read and write in Nepal's Devnargi script. Kanche had none of these talents as when she had grown up in Namche Bazar, where she was born, there had been no village school.

I now realised that Kanche's lodge in Pheriche, even though it bordered on the Kleenex Trail, was a marginal activity at best, open to trekkers only during the height of the high seasons, in spring and autumn, eight weeks at most and then it was lucky to receive the overflow from Pheriche's four other lodges, all larger and better equipped. A bed was 5 rupees a night, milk tea 4 rupees, *dal bhat* maybe 10 rupees — the prices were pretty close to cost if you took into account the overhead, mainly firewood which had to be bought from a wood merchant down the valley.

Now that his climbing days were over, what Gyalzen did for work I was not entirely sure. He was a good cook and an experienced sirdar — of the 'old generation', Bharat Parajuli had pointed out — but trekking agencies tended to hire their staff at Kathmandu as more and more Sherpas gravitated towards the capital. So Gyalzen's trekking work was occasional, and apparently very occasional at that.

I sincerely doubted that the rest of the village knew about Gyalzen's plight. He and Kanche were too proud and status-conscious to let that one out of the bag. Their financial problems were a matter of internal concern, to be worked out within the family through discussion, resolution and hard work. But they did share what little they had with a woman in her early fifties who had a young daughter of about seven and no husband. The woman was hard of hearing and seldom spoke; her eyes were full of sadness and suffering. Her clothes were old and often-mended, but she was neat, unobtrusive, and helped Kanche with household chores. Her daughter, not much larger than Tsheden, with huge round eyes, was cute as a button and, like her mother, almost silent, with a small voice, but perfect hearing and keenly interested in everything that went on. Kanche gave the mother old clothes, for which the woman was grateful, and fed the two of them whenever they came to the house.

Tsheden at all times demanded the centre of attention. I supposed the transition from youngest to next youngest was difficult for her. She could no longer sup at her mother's breast, nor could she command the full attention of her papoo who was now ever so adoring of Mingma Ramu. She had devised all sorts of tricks for focusing attention on herself. And if her tricks didn't work she would wallop brother or sister, mother or father, or break into a foot-stomping tantrum. But that night Tsheden was quiet. She had fever and there was a discharge from her ear.

Saturday, 19 December, was a rest day for everyone except Kanche. For her it was market day. In Sherpa country this obviously does not mean getting into the family station wagon and driving to the nearest supermarket. She got up in the pitch dark at 4 a.m., saddled the yaks, and drove them five hours down to Namche, completed her shopping, and drove them (fully loaded) six hours back up the valley, arriving home at 7:30 p.m., in the dark, exhausted. She had bought tiny shoes for Tsheden Dolma, a carton of salt biscuits, a box of Rara noodles, yak cheese at 200 rupees a kilogram, a shank of yak meat, several kilograms each of carrots, sugar, rice, flour, a jerry-can of kerosene, peanuts in the shell and candies for the children, a score of eggs, a few dozen suntallas and a half-dozen bananas. There was such excitement. She distributed the peanuts and candies. Father got a portion of peanuts.

No sooner was she seated on the floor in front of the fire with a glass of warm chang than she had to bare a breast and feed Mingma Ramu who had been angelic all day. As there are no prepared baby foods on the shelves of shops in Nanche, Sherpa babies when they are old enough to digest tender foods have their diets supplemented by mouth-to-mouth feeding. After chewing bits of potato, rice, tsampa or whatever, the mother spits the morsels into the baby's mouth. The whole family joins in, helping feed the baby in this manner, and so all day Yangtse, Kunga and Pasang Tshering had taken turns feeding Mingma Ramu.

Whilte Kanche was breast feeding Mingma Ramu, I winked at Kunga and nodded for him to come over. He sidled up to me and, covertly, I emptied my share of peanuts into his hands. He beamed and went away. Ten minutes later he was back. He winked and pressed into my hand a fistful of shelled peanuts. Although not a word had been exchanged between us, I had made a friend.

Since the pujah we had seen little of Uncle Pasang and I sensed tension was in the air. We were preparing to return to Shanjo on Sunday, 20 December, and Gyalzen announced that Pasang was not coming. Pasang's wife had demanded that he tend to the yaks and cut wood for the winter. Kanche was disappointed with Pasang. She was a formidable person; a most generous hostess, but also most demanding. She insisted that Pasang return Gerard's old sleeping bag which I had given him at Donag Tsho to supplement his very tired old bag because he had been freezing at night.

I explained to Kanche the circumstances under which Uncle Pasang had come by the bag and why I didn't think he should be asked to return it. She thought about this for a moment, then rendered her decision. Pasang would be asked (nay, ordered) to return the bag for the duration of the 'expedition' and it would be given back to him upon my departure. We needed it, she said, if the family was to accompany me to Shanjo.

Uncle Pasang was called on the carpet on Sunday morning and told to return the sleeping bag. Kanche harangued him mercilessly and I felt sorry for him. Gyalzen paid him off at the rates set by Bharat Parajuli. Pasang returned a short time later with the sleeping bag and made a point of handing it to me. 'Goodbye, Father,' he said. I hoped this was only a passing contretemps as I had become attached to the dependable, gentle-spoken Pasang.

We decided to return to Shanjo the next day. Unfortunately, our second visit was uneventful. There were no signs of new tracks other than those left by the wandering fox and family of mouse hares. We tried Bibre, half a kilometre across the Imja valley from Shanjo, where the Australian trekking guide Gary Scott said he had seen two yetis. Above Bibre we found old tracks that kept the Gary Scott story alive, and dozens of impeyan pheasants. From the top of one 5,450-metre mountain, we gazed across at Kongma La, the pass behind Pokalde that leads westward to Lobuche, and saw several lines of footsteps, some leading nowhere, others heading towards the pass.

'Tourists,' Gyalzen assured me. But I was not so sure. In any event it was too late in the day to reach them and once back at Shanjo it would have required another day to climb to Kongma La, four kilometres distant and at an altitude of 5,535 metres (18,155 ft). With hindsight, I should have made the effort to inspect them, especially after seeing tracks six weeks later on Khumbu Glacier coming from the west side of Kongma La, but as I had so often expended what seemed like vast amounts of energy at these altitudes only to be disappointed, I let the opportunity pass.

Gyalzen suggested that in two days we attempt Island Peak. Island Peak is 6,189 metres (20,305 ft) high, which was 1,720 metres (5,642 ft) higher than our campsite at Shanjo, and ten kilometres east of us, between the two glaciers of Lhotse and Lhotse Shar, which join at its southern base to form the Imja Glacier. A solid white pyramid, its summital ridge is a prolongation of Lhotse Shar's south-east buttress.

The next morning, while lying in my sleeping bag, I and several million other listeners on the Indian subcontinent learned from Voice of America about the existence in the United States of something called the Ho-Ho Academy which teaches old folk how to play the role of Mr and Mrs Santa Claus. Children from all over America could call a toll-free number and get one of the Ho-Ho trainees on the phone. The callers could then tell the Ho-Ho trainees impersonating Mr and Mrs Claus what they wanted for

Christmas. Now, wasn't that nice, I thought.

This ten-minute radio feature included an interview with the academy's director who underlined the important social role that the academy filled in serving the young folk of America while at the same time giving the country's retirees a useful function. A little old lady was interviewed about her work as one of the Mrs Clauses. This snippet from the American Dream was supposed to fascinate the average Indian, a Hindu who couldn't care less about Santa Claus, assuming he or she knew who Santa was. Neither Gyalzen nor Kanche knew about Mr and Mrs Claus though they did know that tourists from the West had a big pujah called Christmas. I wondered about the asshole in Washington, earning a salary in the upper five-figure bracket, who had programmed this glimpse of American tribal custom for transmission to Asia while a constitutional crisis was bleeding Bangladesh to death, an Indian expeditionary force was fighting a bloody war against Tamil separatists in Sri Lanka, and the worst tragedy in maritime history had catapulted the Philippines into pre-Christmas mourning.

Christmas Eve was the loneliest I had ever spent. The recurring radio theme broadcast from London and Washington was the gaudy over-commercialization surrounding the celebration of Christ's birth. My thoughts were with my children. I missed them more than I would let myself admit. I prayed for them, hoped they were happy and well, and I prayed for their mother, wishing that she too, was happier than in years past. And finally I willed myself to sleep while listening to a BBC orchestra play the Nutcracker Suite.

On Christmas morning Kanche, who had arrived the day before with Mingma Ramu from Pangpoche, made hot raksi eggnog with rancid yak butter and stood over me till I had consumed two cups. Then she gave me some crystal sugar that she had bought at Namche market. 'Very good Sherpa medicine,' she said, explaining that if I sucked it on the way to the top it would stop me coughing.

Gyalzen and I left for Chhukhung around 10:30 a.m. The weather was perfect: bright sunshine, no wind. Chhukhung was desolate: a battlefield of glaciers from where we began the long walk up the giant moraine between the Lhotse and Imja glaciers. I was still burping up Kanche's eggnog. By then it had formed a knot in my stomach and made my legs feel spidery. I complained to Gyalzen that it was not serious to make me drink rancid eggnog before setting out on a hard climb.

'How many cups?' he asked.

I said I had drunk two.

'Two no good,' he replied seriously. 'If you had taken five cups you would feel strong as a lion.'

We arrived at our base-camp site around 3:30 p.m. We were at 5,100 metres (16,732 ft) and about to lose the sun. Very quickly Gyalzen set up the tent on a grassy platform bare of snow, fetched water from a spring 200 metres away before it froze for the night and had a pot of tea brewing on a fire of yak dung and twigs that he had brought with him in a plastic bag from Shanjo.

Less than 100 metres below us was the base camp where Rita Ann Scholtens, 35, Graham Sorby, 27, their sirdar Gopal Tamang, and his brother, the cook Suk Man Tamang, were killed in the avalanche caused by the Black Monday storm of 19 October. It was bad luck. The avalanche had broken off from the shoulder above them while they slept. Their mangled tents, some clothes and other gear could still be seen in the rubble of the snow. John Scholtens had built a rock altar with the names of the four victims engraved in stone. It was sad to visit them on Christmas day like that. The place had a feeling of tragedy about it.

Opposite was the junction of the Amphu Labtsa and Imja glaciers. The backdrop to the east was a vast amphitheatre of ice and snow of truly Himalayan proportions while lurking unseen behind Island Peak was Chomolungma. As soon as the sun went down, the mercury plunged to minus 35 degrees centigrade. Gyalzen served me Rara noodles in my sleeping bag. A serac broke off under Island Peak's south summit with a tremendous roar. We wondered whether it was going to cascade out of the bowl above us and dust us away. The glaciers grumbled into the night. Neither of us slept more than a few winks.

Gyalzen started making coffee at 3:30 a.m. and filled the thermos with hot tea as I got ready. By 5 a.m. we were underway by the light of our headlamps. At 6 a.m. I had frostbite on one toe. It was light enough to turn off the headlamps. At 6:35 a.m. we reached Camp I and kept climbing, entering a steep corridor filled with unevenly frozen snow. We exited the corridor onto a rocky ridge at Camp II and attacked the mountain's main rampart; a sinewy spine on which every rock seemed to move. The upper half of the spine was dressed in peeling slate, with a sharp drop of several hundred metres on the left, less on the right. We were unroped.

At 5,700 metres we climbed onto the mountain's snow cap and stopped in the first sunlight for breakfast of tea and biscuits while we put on our crampons. It was 8 a.m. We set out across the glacier for the summit ridge through seracs and around crevasses so gaping that whole houses could have been swallowed in them. We reached the base of the summital ridge one hour later to begin the most treacherous part of the climb: a 180-metre wall of mixed slate, ice and rotten snow.

We crossed the rimaye without problem and climbed through a slate cleft with our crabs on, unroped, then negotiated an ice-filled chimney to a wall of snow above it. Relief. The climbing, while semi-vertical, became easier. The exit onto the summital ridge was across dabbled snow frozen as hard as ice. For this we roped ourselves together. Once on the saddle of the ridge all that remained was a 200-metre stroll to the summit, rising in two steep pitches. Most of the time, because of a cornice overhanging the east face, I had to walk on the west side of the snow crest. Gyalzen was behind me. One false step would have sent us slithering down the snow-clad west face to the Lhotse glacier 1,000 metres (2,808 ft) below.

We reached the summit at 10:30 a.m. It had taken five and a half hours to negotiate a vertical ascent of 1,150 metres. It was Boxing Day. The sky was a deep velvet blue, the snow pristine and there was not a breath of wind, nor cloud anywhere in sight. Lhotse Shar looked so close it seemed we could touch it and the climb up its south-east ridge to a summit 2,200 metres above us looked, from our vantage point, like a piece of cake.

We could see clear down the network of valleys to Namche Bazar. Everest was hidden from our view by the Lhotse-Nuptse massif. The temperature was more akin to the Cote d'Azur than the attic under the roof of the world. We had a snack, snapped pictures, tossed rice to the gods and, mindful that the sun was rotting the snow, started our descent after a half-hour rest.

Gyalzen left me below Camp I so that he could have hot noodles waiting when I stumbled in to base camp at 2:45 p.m. During our absence the tent, mysteriously, had been ripped in two places. Were the rips caused by a tahr who mistook its brown dome for a giant moss? Or was it the yeti?[2] Gyalzen, I'm sure, thought it might have been the restless spirits of the avalanche victims. In no way did he want to spend another night there when the only thing I wanted was to crawl into my sleeping bag and collapse.

He packed while I ate and rested. We broke camp one hour later, arriving at Island Peak Lodge in Chhukhung in the dark at 6:10 p.m. Our day had begun by the light of our headlamps and that was how it finished. We had been on the go for thirteen hours at altitudes higher than Mont Blanc. I was too tired to eat but forced down two bowls of lentil soup. Gyalzen rolled out my sleeping bag. Other than the lodge owner and her two children, we were the only people in the lodge. Although I shivered myself to sleep, I slept like a dead man.

Gyalzen brought me tea in bed at 8.30 a.m. We packed quickly and left for Shanjo where Kanche, Pasang Tshering and Mingma Ramu were waiting. We sat in the sun, ate tsampa, drank tea and joked about the raksi eggnog. We decided to leave for Pangpoche immediately.

---

[2]We figured out some weeks later that the rips were made by yellow-billed choughs attempting to get at the food we had left inside.

Chapter 14

The Blessing

Kanche came into the lhang at 7 a.m. with a cup of hot chang. As she handed it to me she announced that everybody was to have a sponge bath that morning and put on their best clothes. We had been invited to the village lama's house for lunch.

We brought our lunch with us. Gyalzen also gave me a *khata*, the symbolic white scarf of peace and friendship, with a 50-rupee note folded into it which I was to offer to the senior lama who was visiting the village. He was, Gyalzen explained, a noted teacher from Tengboche who had arrived with his thawas to hold two months of classes in Pangpoche. It was a field trip for the thawas, learning about the life and teachings of Sanga Dorje at the very place where the venerated lama had performed most of his meditating and miracles. Gyalzen, dressed in the traditional black chuba, carried with him a small bag of rice and another of millet.

The house was about 50 metres above the gompa, at the top of the village. It appeared to be an ordinary Sherpa house, with stables downstairs and a long room with kitchen upstairs. One of the students was cooking in the kitchen and a deformed idiot child cowered in a basket by the window. A separate room had been built for the teaching lama off the terrace at the far side of the kitchen.

We ate our lunch with the thawa in the long room, the village lama being absent and the senior teaching lama remaining in his room off the terrace. Gyalzen wanted the teaching lama's blessing, which the senior lama had apparently agreed to bestow before the afternoon classes resumed. So as not to be late, Gyalzen hurried through the mountain of *dal bhat* that Kanche had ladled onto his plate. I was left alone with the student thawa and the idiot child to finish my heaping portion of half-fermented coleslaw.

After ten minutes, Kanche came to get me. The senior lama wanted me to attend the ritual blessing. I brought the khata with me and laid it beside him. He thanked me and indicated I should sit on a settee that was to his right. Gyalzen was seated cross-legged on the floor directly in front of the lama. Behind him were the five children and Kanche.

On a table to the left of the lama were the ingredients of Everlasting Life, and a vajra, or dorje, representing the power of the lightning bolt. The other items included a peacock quill in a well, a bell with a dorje handle, a vase of holy water tinged with saffron, a saucer containing a sweet-tasting mustard-like substance, a hand rattle-drum, a bowl of chocolate cherries, representing the pills of life, a carved wreath with a horse's head and a hoof-like base, and a wand with cosmic-coloured gauze tied to it.

The senior lama had a monumental sinus problem. He was dressed in orange and crimson robes. He seemed a big, hearty fellow, and sat cross-legged on his settee in front of a Chinese lacquered table. He resumed the ceremony, sprinkling holy water on Gyalzen's offerings of rice and millet which had been emptied into bowls and placed on the Chinese table in front of him. He recited from the tantras in a running chant, invoking Tamdin, king of the protective demons. He then motioned me to stretch a cupped hand towards him, into which he dipped some holy water with the peacock quill and I was required to lap it up. The others had already received their helping before I was admitted to the ceremony, whose principal benefit, Gyalzen later told me, was to provide protection against renewed illness.

Gyalzen's father and mother arrived with their own bag of rice, accompanied by two friends, one of them carrying a little bowl into which, in addition to her portion of the blessed water and pills of life, went another portion for a bedridden relative who was to receive the blessing by proxy. The senior lama occasionally interrupted himself to blow his nose into a towel. Gyalzen was blessed in triple measure, the rest of us once, each receiving a chocolate cherry, some of the saffron water, a dash of this and a sprinkle of that. When the ceremony was finished we were motioned forward one at a time to receive a red string which he draped around our necks.

A short break followed while the room was cleared and the windows giving onto the terrace opened, after which about fifteen thawas in their early to mid twenties arrived for their afternoon lesson. They joked and pulled gags on each other. Two had Walkman cassette players, and most had impressive wrist watches. The reading from the holy texts which they were to learn that day lasted almost three hours. Throughout the recital, Gyalzen, his parents, Kanche and I sat on the terrace, among minute potatoes and corn kernels spread on mats to dry. Several of the students nodded off to sleep in the warm afternoon sun.

While the senior lama was reading to his students, the dollar went on the skids again after a three-week rally, and this in spite of concerted Japanese and European central bank intervention. Gold in Zurich rose to $490 an ounce. Stock markets trembled and share prices went sharply lower. American retailers announced that Christmas sales had been disappointing, up only 3 per cent over last year, which was less than the rate of inflation. A slowdown in consumer spending was forecast for 1988.

Overnight a strong wind rose and then subsided, leaving a layer of grit over everything. Tuesday 29 December, was a raksi-making day. The tasting of the new production that evening contributed to a general feeling of euphoria. A note arrived from Bijaya Kattel to say that he had returned to Phortse and was hoping to see us.

Chapter 15

Saturday Market

'Father little headache?' Gyalzen asked when he entered the lhang with what I hoped was tea. I admitted that I had a slight headache. 'Then little raksi,' he said, handing me a warm glass of Kanche's potent brew. 'Best medicine,' he promised.

In the long room a mountain of shredded potatoes was heaped on a plastic sheet in the middle of the floor, the product of last evening's industry, still waiting to be distilled. Kanche was laying in a supply of strong beverage for Lohsar, the Sherpa New Year, which was three weeks away. I had decided to visit Bijaya Kattel in Phortse. I assured Gyalzen, as he had things to do at home, that I could make the journey on my own.

We had decided that our next sortie would be to Konar and Macherma, but before setting out we had to make further purchases in Namche. Gyalzen agreed to meet me in Phortse the next day so that we could continue to Namche together for the Saturday market. When I arrived in Phortse I found Bijaya's door padlocked. The neighbour told me he had gone to Tengboche that morning with his son and nephew. There was no point in returning to Pangpoche, so I took the trail down to the bridge over the Imja and climbed the ridge to Tengboche, arriving there just after 4 p.m. I met Pemba, the Thyanboche Lodge manager, who told me that Bijaya had an audience with the high lama.

Pemba, an engaging fellow — I suspected he was a police

informer — went to the high lama's private quarters and inquired when the audience would finish. He returned and told me that the high lama had asked that I attend.

It was now dark and Lama Jangpo's private audience room was lit by a kerosene lamp. He was seated in his robes, looking very relaxed but inscrutable, on a settee at the far end of the room which was no more than three giant steps long and one and a half wide. In front of him was a red lacquered table on which sat a silver-hatted tea bowl, and behind him on the wall were three Tibetan scroll paintings.

Lama Jangpo bid me take a seat beside Bijaya, who explained in Nepali that I was looking for the yeti. The lama's eyes sparkled at the news, of which he was already aware. He told Bijaya in Nepali, for he spoke only a few words of English but listened to the Hindi service of the BBC every night, that he had never seen a yeti. As a boy, however, he remembered being shown the footprints of one that had circled the monastery counter-clockwise — that is, opposite to the way devout Buddhists should circumambulate religious sites — early one wintry morning while everyone slept.

The region of Tengboche used to be popular yeti country but none had been seen around the monastery for more than thirty-five years. The last time, in fact, was during the winter of 1951, when a yeti descended Kangtaiga ridge. Lama Jangpo was away in Tibet at the time, studying, but had been told about the incident. The same yeti had been seen three years before at the very same spot on the ridge. On both occasions the monks brought out their musical instruments and kicked up such cacophony with trumpets, drums, cymbals and conch shells that it ambled off into the rhododendron brush.

Charles Stoner, a member of the 1954 *Daily Mail* expedition, spoke to Lama Jangpo's regent at that time, Abbot Nawang, about the existence of the yeti and described the conversation in his book *The Sherpa and the Snowman* (Hollis & Carter, London 1955). 'Queries as to there being more than one kind [of yeti] brought the answer that Abbot

Nawang knew the large chuti quite well. He had several times seen chutis when journeying through Tibet, but did not think they were ever found in Sherpa country. It was very like the black bear of the Solu country lower down, only larger and with thick reddish fur. It was a great scourge to livestock, and dangerous to man. It is best to keep out of the way if you meet one.

'The much smaller mitey . . . is the kind seen in Sherpa country. It walks upright, and has the habit of always taking the shortest and most direct route when on the move. Nawang himself had twice seen one. The second time was three years ago, when it appeared in cold, snowy weather, and the monks watched if plodding down towards the monastery. It stopped to sit on a rock and scratch itself and evidently was watching them. They were so frightened that everyone turned out, blowing conch shells and banging drums, to drive it away. At this it jumped off the rock and disappeared . . .'

Lama Jangpo had been well versed in the lore of the yeti since his earliest childhood. As a thawa novice he had studied at Trashilumpo Monastery in Tibet's second largest city of Shigatse. On cold winter evenings in the prayer room the monks would wrap him in a dark yellowish-brown pelt which he remembered as being very warm. The monks told him it was a yeti hide. Trashilumpo, which signifies 'full of benediction,' was the headquarters of the Panchen Lama until destroyed by the Chinese after the 1959 Tibetan uprising.

He also said that the yeti was mentioned in Tibetan medical texts. I asked which ones. He replied that he had never seen the texts but his teachers had told him about them.

Bijaya mentioned the several yeti artifacts that existed in Khumbu, such as the skeleton hand at Pangpoche. Lama Jangpo said he doubted it was really a yeti hand but probably had been severed from a thief as punishment for his misdeeds, as such in times past was the sanction for stealing. He thought some people probably attached mystical powers to the hand, thinking it would make them rich.

With this he related a macabre tale that had occurred a good many years back, after a forest fire in Kharikhola valley had killed two herdsmen trapped at a rock camp. He said an enterprising American happened to be in the area at the time and came across their charred remains. He sectioned one of the dead herdsmen's hands, the flesh having been burned off it, then visited Pangpoche gompa to see how the skeletal hand there was presented and made an exact copy of the satin-lined box, placing his 'treasure' in it. The American took the package to Tengboche, convinced that he might have an eager buyer as he knew the monastery to be poor in yeti relics.

Lama Jangpo, however, said he knew exactly what the American was up to as he had received a report about his doings from people who had followed him from Kharikhola. He said the American was sent packing, politely of course, and he never heard whether this ghoulish adventurer found a buyer for the 'yeti' relic.

While Bijaya had dinner with the high lama, who had business to discuss with him, I went to Gompa Lodge and met Andy Teare, a zoo vet with the Smithsonian Institution. Andy had come to Khumba to see how Bijaya was getting on with his musk deer project, but was leaving in the morning to return to Washington. We speculated about the yeti for a while and then retired to bed in the lodge's dormitory which was full of Japanese trekkers.

Incredibly, the year had suddenly come to a close and next morning, Thursday, 31 December, a motley crew stood in the sun on Tengboche common waiting for a party to happen. For Bijaya there would be no party. His nephew had come down with altitude sickness and, vomiting, had to be evacuated to Lukla. We agreed to meet again in Phortse the following week.

But those preparing to celebrate the Julian New Year included Todd Hoffman, a real estate developer from Hartford, Connecticut. Hoffman, 33, had sold his Porsche agency the year before and was now possessed of itchy feet. Before leaving the United States for a year-end trip to Kathmandu he informed me he had sold more than $1 million in securities so as to be sheltered during his absence from further market vagaries. A Japanese woman, who had climbed in Chamonix with her husband, drifted over and talked to us. The sunshine was glorious, the view up the valley, beyond Pangpoche to Nupste, Lhotse and Everest was ever spectacular and the goings-on in front of the monastery always entertaining.

A Nepalese Army climbing team preparing for a 1988 Everest tripartite expedition, strolled up the ridge from Deboche, accompanied by Gillian, a baggy-trousered 26-year-old groupie from Illinois. Gillian had given up a job as a buyer for a chain of American department stores to travel, alone, through Asia. She had already crossed China, come down the silk road into Pakistan without being overly molested, and now was contemplating the Himalayas. She was quiet, observant, and proved to be good company.

The Army team was led by a barrel-chested Gurung colonel who carried a stout walking pole, and included a bald-headed doctor, a spare major and a laughing lieutenant, one or two subalterns and a bevy of orderlies. There was also Steve, a Canadian trekker, who had been with the Army group at Lobuche. The only missing person was Gyalzen. I assumed that in Phortse they would tell him I had gone to Tengboche and that eventually he would turn up. Meanwhile, I relaxed in the sun and watched the party develop.

The centre of attraction was the colonel. Court was being held around him. I had developed a craving for coconut biscuits, of which the Nepali Biscuit Company make a very fine variety. Fortunately the trekkers' shop at Gompa Lodge stocked them. I exchanged one of my coconut biscuits for a Nepalese Army glucose one and a conversation developed. We chatted for a while until it dawned on me from things that were said I was talking to the former Nepali military attaché in London.

'I think we have a common friend,' I told him.

'Who's that?' he asked. He looked doubtful.

'Nick Claxton,' I replied. Nick was a British TV producer who won an Emmy for his documentary on Ethiopian famine called *Cry Ethiopia, Cry*, which started the whole BandAid movement.

The colonel looked at me for a moment. 'You're Hutch!' he finally said. 'What are you doing here?'

'Looking for the yeti,' I said.

'That's funny. Hemanta called me before I left Kathmandu and said the King Mahendra Trust wanted to organize a yeti expedition. He wants me to lead it. We had a meeting at his office.'

'Oh?' I said.

'Yes, the prince is very keen on the idea,' he added.

Colonel C.B. Gurung, in the estimation of some people, is married to the most beautiful woman in Nepal, herself a television presenter and producer of environmental documentaries. C.B. was the commanding officer of the Kesang military training school near Jomoson. He had attended the Indian Military Academy in Dehra Dun, which was next door to the Indian Forestry College, where Hemanta Mishra had studied, and the two became not only friends but the joint terrors of Kathmandu society during the 1970s. He stood about two metres tall and towered over Gyalzen when finally the Pangpoche pundit arrived from Phortse in the early afternoon.

While we were talking on the common, a white Puma helicopter from the Royal Flight landed behind the third lodge amid clouds of dust. It was carrying a four-man television crew led by an Austrian, Kurt Diemburger, wearing a red wet suit. Diemburger, then 55, had climbed Everest in 1978 and K-2 in 1986. He was one of seven climbers caught in a week-long storm at 8,000 metres when descending the Pakistani peak. Only two survived.

The nucleus for the party was now complete. Gyalzen, I could tell, felt overpowered by the gathering of celebrities, and so we retired to Gompa Lodge for a glass of chang with Thawa Nang Wasser, the honourable lodge keeper. An hour later the laughing lieutenant found us and announced that the colonel requested our presence at a New Year party he was hosting that evening. I looked at Gyalzen. I knew he had important business at the bank in Namche in the morning. The bank closed at midday for the Nepali weekend and Gyalzen was determined to get there well before then. He looked at me and said, 'Father, we go Phunki.' With a heavy heart, I had to tell the lieutenant that we were leaving in a little while but looked forward to meeting them at the market in Namche.

The teahouse at Phunki, in front of a line of water-powered prayer wheels, is owned by Gyalzen's friend, Nang Lobsang, and his wife. It has a small dormitory attached to it. Nang Lobsang and Gyalzen had been student thawas together and I could imagine they made a mischievous pair. Nang Lobsang's wife is reputed to be the best chang and raksi maker in Khumbu and I can attest that this would seem to be correct. But because their homemade brews were in much demand by the local clientele they would not often serve these beverages to tourists, who have no appreciation of such things. Gyalzen was eager to try the latest barrel of chang and I was invited to keep him company. While we sipped from our glasses, Nang Lobsang came over and asked to borrow the pair of binoculars I carried around my neck. He took them to the door and focused on the mountain opposite, one of the progeny of Kangtaiga. He stood there for the longest time, studying the upper reaches of the mountain through the glasses. I wondered what on earth he could be looking for. Finally he came back, shaking his head, and announced: 'Yeti sleeping.'

Before dinner we were joined by a gentleman from Osaka, his guide and a porter. Afterwards we shared a pot of raksi and a package of coconut biscuits, then retired to bed. Listening to the BBC I learned that the Dow Jones closed the year at 1939, up only 2.25 percent over the twelve month period. The US inflation rate for 1987 was 4.5 percent, so investors who did no better than the Dow Jones were out of pocket. The dollar closed at a new 40-year low against the yen, pound sterling, deutschmark and Swiss franc.

When I awoke on New Year's Day there was a weasel on the bunk next to me. During the night he had been having a snack of Tibetan dried lamb, a shank of which Nang Lobsang kept in a box by the door. The weasel seemed pretty pleased with himself and was not unduly concerned that I wanted to get out of my sleeping bag. Gyalzen showed Nang Lobsang the gnawed shank and the teahouse owner became greatly concerned. When we left he was plotting revenge. Todd Hoffman, already on his way back to Lukla, joined us on the trail to Namche. Todd had to fly home to Hartford so that he could re-invest his capital now that the market was rebounding. He said we had missed 'one helluva good party'.

In Namche, we checked into Tramserku Lodge and I had my first hot shower in two months while Gyalzen went to the bank to conduct his business. It took him the rest of the morning. The bank was located in a two-storey house at the far end of the east-west transversal path. I waited in the sun under Thawa Lodge for him to finish.

'What took so long?' I asked when he saw me sitting on a rock at the side of the path. He looked greatly pleased.

He suggested we have a cup of chang at Kalapattar Lodge.

Upstairs at the lodge, with a view over the Namche common, where the seeing-eye stupa and a wall of mani-stones are located, he told me his story of economic woe. He said it had been very difficult after I left the previous spring, without much work about. I had understood by then that lack of cash had been responsible for Pasang Tshering's withdrawal from the monastery school. Ever since, the young antelope, too old for the village primary school and opposed to going to the secondary school at Khumjung because he didn't think it up to his standard, had been helping his parents at home.

Gyalzen had been too proud to tell me this. When a few weeks previously I had asked why Pasang Tshering was no longer at the monastery school he told me ingenuously that someone at the monastery was a bit *go*, meaning crazy, and had taken a dislike to Pasang Tshering, playing nasty tricks on him, like ripping his sleeping bag and cutting holes in his down jacket. So Gyalzen said he and Kanche had decided to take him out of the school and bring him home.

Anyway, to get through last winter, Gyalzen had borrowed 6,000 rupees ($240) from the bank, using as collateral a gold hair braid and gold ring belonging to Kanche. The rate of interest was 18 percent. With the funds I had given him, he had now been able to pay off the loan. What had taken so long, he said, was that the bank manager had been unable to find the guard with the big shotgun who kept the keys to the vault. Gyalzen removed from his pack a bundle and unwrapped it to show me the handsomely crafted gold jewelry.

'From Tibet,' he said. 'Very old. Kanche will be happy.' I could have sworn there were tears in his eyes and I lowered mine to look at the gold braid so as not to embarrass him.

That afternoon I suggested a visit to Thawa Lodge, hoping to show Gyalzen the mummified yeti foot, which belonged to the lodge owner and which Gyalzen had not yet seen. The owner, Tenzing Tsherpin, had also been one of Gyalzen's classmates at the monastery school. I had first seen the foot in March 1986 when I came through Namche with Gyalzen's brother-in-law, Lobsang Tshering, from Bodnath.

The foot's history was interesting. Tenzing Tsherpin said he had been vexed when the 1960 Hillary expedition declared that the yeti did not exist. Because the yeti is the guardian of the secret abodes of the gods he knew that it had to exist, though he had no proof. Several years later a Tibetan trader came to Namche carrying this foot wrapped in gauze. It was after the worst sacrileges had been committed by Chinese troops who in an orgy of fury following the 1959 uprising destroyed thousands of Tibetan monasteries, gompas and other religious monuments. The trader said it was a yeti foot which he had taken from one of the ransacked monasteries.

Tenzing Tsherpin purchased the foot from the Tibetan and kept it, a secret, in a drawer under the pulpit in his private chapel. Now, he said, he had proof that the yeti existed.

He took us to the old house behind the lodge and we

climbed a set of stairs to the unused long room. The house was much larger than Gyalzen's with a separate wing for the lhang which was lit by two large windows. A lot of money had been spent decorating the lhang with elaborate murals featuring religious motifs. They were brilliantly executed and lovingly maintained. Tenzing Tsherpin was obviously a man of substance.

From a drawer under the pulpit, to the left of the altar, he removed an object wrapped in old gauze. He placed it on a table by one of the windows and unravelled the mummified foot. As during my previous visit, he refused to allow me to photograph it. He maintained that Gyalzen, Lobsang Tshering and myself were the only people outside his family to know of the foot's existence and he only agreed to show it to us because his wife, Gyalzen's wife and Lobsang Tshering's wife were cousins. Furthermore, he said, Lobsang Tshering, may the gods bless him, had described me as a great authority on the forces of mysticism.

It was a left foot, measuring 18.5 centimetres (7.3 inches) long (without the heel), 9 centimetres (3.5 inches) wide, with five more or less equal toes and hardly any bridge but a good instep. From what normally would be considered the big toe, and it was marginally bigger than the others, protruded a 5.3 centimetre long (2.1 inch) curved toe nail. It was yellowish black to brownish grey. The two toes next to it also had nails that were split off close to the skin. The last two toes had holes where the nails had once been. It was covered with coarse greyish skin with, in places, traces of reddish-brown hair.

Andy Teare had suggested I try to borrow the foot and have it X-rayed at Kunde hospital above Namche. But Tenzing Tsherpin said this was out of the question, nor would he let me show it to Bijaya Kattel. I could, however, buy it for $5,000: otherwise I was left to admire it for another few minutes and that was that.

The Tibetans refer to a *mi-go* in their literature and records of sightings. A *mi-go* is sometimes confused with a yeti. *Mi* means man; *go* in Sherpa-ka means crazy, and in

Tibetan means wild. So a *mi-go* is a wildman or *almas.* In my view, Tenzing Tsherpin's foot is authentic and probably once belonged to a *mi-go.* The 1958 Slick expedition to the upper Arum valley claimed to find tracks that corresponded to a *mi-go's* foot. One of the members of that expedition, Peter Byrne, had also claimed to have seen similar tracks in forested areas of Sikkim, and it might explain tracks that I would later see in Macherma valley.

Tenzing Tsherpin carefully wrapped the foot in its gauze and replaced it under the pulpit. Back in the long room I admired the dimensions of the beams. The house was more than one hundred years old, he said. The width of the main beam was easily 60 to 70 centimetres. I asked where it had come from. 'Namche,' he said. 'But trees like that don't grow here anymore.'

The colonel's team had arrived by then at Trekkers Inn and we went to have drinks with them. A boisterous Gurung police officer from the Thame checkpoint, who had been drinking most of the afternoon with the Namche immigration officer, was in the lodge kitchen, as was the lodge owner's 18-year-old son. The son was back from his first term at college in Kathmandu and seemed very excitable. The police officer irritated him and finally the Sherpa boy told the Gurung to shut up. To rationalize his aggression towards the police officer he came over to us and, so that the officer could hear, complained that when petty officials from Kathmandu arrived in Khumbu they thought they owned the place. The police officer lunged at him, threatening to 'wring his neck', and had to be restrained.

Among the tourists staying at Trekkers Inn was a young man from Illinois by the name of Gary who took me aside. 'Are you really looking for the yeti?' he asked.

After assuring him that I was the person he had been told about, he confided that he had come from Macherma where on 30 December he had found fresh yeti tracks crossing the valley. He described them and said he had walked beside them for some distance. The stride was long and the foot imprint large enough to step into. He said he had seen somebody else walking beside them and photographing them. He thought I should know. I thanked him.

Kanche, Kunga and Mingma Ramu arrived with the yaks in the late afternoon. They were among the swarm of people pouring into the village for the weekly market. Every Friday the village of Namche begins to take on a gala air. Government officials, especially, congregate early Saturday morning on the top terrace of the marketplace and survey the scene, happy to find each other and compare notes on the rigours of life away from Kathmandu. They might include the district officer, the local army commander, the engineer in charge of the Thame hydro project, the director of the yak-breeding station above Namche, the chief of police, the park warden, the head customs inspector, the immigration officer, and any visiting dignitaries or expedition liaison officers.

The merchants who sell their goods in the marketplace are taxed a small sum, half of which goes to the village council, such as it is, the other half being paid into the government coffers. As the roadhead is at Jiri, 154 kilometres away, the merchants must carry their goods to Namche each week, an enterprise in which whole families will sometimes participate, particularly wives and daughters, as throughout the Third World women are the most common and cheapest beasts of burden.

Up to 500,000 rupees ($20,000) change hands on a Saturday in Namche and leave town that same afternoon or the following morning. For the locals this was a stupendous amount, though at any fashionable restaurant in New York it represented the trade of a workaday lunchtime.

While Gyalzen and Kanche did the shopping, I joined C.B. and the others at the top of the marketplace. The four rows of terraces were bursting with people, dogs and crows, the latter hoping to pick up morsels of raw yak meat which low-caste butchers were cutting up in front of their clients, mostly lodge owners and trek operators. The unwanted morsels were thrown into the pathway or left on a terrace wall. Gillian was buying all the suntallahs she could carry at one rupee apiece. At mid-morning we were abducted by

Yanchin, owner of Above The Cloud Lodge at Lobuche, for steamed momos and chang at her sister's lodge in the centre of Namche. C.B. invited everyone to lunch at Trekkers Inn. It was a moveable party.

When we arrived at Trekkers Inn a family crisis had erupted. The 18-year-old son had made his parents and sister cry by telling them they were uneducated. The kid had learned to drink and chain-smoke while in Kathmandu, which the army doctor said was not normal in Sherpa society. The parents had invested their savings in the boy and were upset by what their toil and money had begot. The doctor, who had begun his medical career as a volunteer practitioner in the slums of Calcutta and had experienced more of the seamier side of humanity than the boy would ever have the privilege of knowing, took him aside and spoke sternly to him. C.B. suspected the kid was on drugs. A recent article in *Rising Nepal*, Kathmandu's daily newspaper, had described drug addiction as a national scourge. The newspaper reported that one in seven households in Kathmandu contained a drug user. A former chief of police and an ADC to the third prince had recently been arrested for trafficking, tried, sentenced to life-long prison terms and their property confiscated as a warning to others that the situation was out of hand.

Around 3 p.m., as we were having lunch with the local notables, smoke started billowing up the hill. A runner arrived to report to the district officer that a bush fire was out of control below the Everest View Teahouse. A careless trekker? No one knew. The Army was mobilized and it took two hours to extinguish the blaze. Khumba by then had several hectares less of tree cover.

Chapter 16

Yeti Marg

Tuesday, 5 January, was overcast and snow seemed assured. We were back at Pangpoche, preparing to leave with Pasang Tshering, Yangtse and the two yaks for the back country of Macherma, which everyone assured us was the likely place to find a yeti.

We spent the night at the home of Gyalzen's cousin in Phortse and next day moved to Konar, camping by the stream at the upper end of the valley. We wanted to explore the far side of Taboche peak and see whether any tracks were on Konar Pass before continuing to Macherma. Four families were at the hamlet when we arrived and their children helped set up our tents, fetch water and organize the kitchen in the ruins of a herder's stone hut.

Next morning Gyalzen and I climbed into the high valley, reaching the terminal moraine of Konar Glacier as the clouds rolled in from the south. It was yak pasture laced with edelweiss all the way up to the glacier. It appeared to be excellent yeti country, with three or four rock camps and plenty of caves. But not a yeti was in sight, nor any trace of one. What depressed me more than anything was that, unless it was the hidden valley behind Kyajo Ri, which we had not yet explored, the high valley of Macherma, or possibly the back country west of Donag Tsho, this area of Khumbu appeared to have no vacant space where a family of yetis could retire with reasonable assurance of not being disturbed. Everything was either yak pasture or trekking territory.

At midday we called it quits. It was cold, with strong

bursts of wind and occasional snow flurries. Pasang Tshering and Yangtse had tea waiting for us back at camp. The wind dropped and that evening it started to snow. Heavy, moist flakes pattered against the kitchen roof. 'We are very lucky,' I told Gyalzen. 'We have good *khana* (food). Lots to drink. And very good fire.'

'But no yeti,' he added, having read my thoughts.

Five centimetres (two inches) of snow fell during the night, the second snowfall in the eleven weeks since I had arrived in Khumbu. But when the sun returned in the morning the snow soon melted. We decided to walk back to Phortse and buy hay for the yaks. Bijaya Kattel had not yet returned, so we spent the morning in his landlord's kitchen drinking chang and listening to yeti stories. The village lama was reciting prayers for the landlord who was preparing to bring home his new bride for the first time. He had given Bijaya notice to move as he needed the quarters. Between mantras, the lama joined in the conversation. He was of the red-hatted orders and therefore not vowed to chastity, which meant he could marry and have children. He mentioned that at the end of the monsoon three years before (1985) his eldest son and some friends were bringing yaks down from Gokyo when, near the second lake at Longponga, they saw a yeti climbing through the rocks unaware that it was being observed.

We asked when a yeti had last been seen at Konar.

The lama didn't need to think for a moment. In April 1986, after a heavy snowstorm, he said. Six or seven 'yak people' saw one preparing to descend into the valley from the high rock camp where we had been the day before. It had waded through snow up to its waist to get to the rock camp and was resting, looking into the valley. When the 'yak people' in the lower pastures saw it they started shouting. The yeti became scared and returned up the mountain, disappearing from sight.

That same spring that yak had been killed at Macherma, by the cliffs at the back of the valley, the landlord's brother remembered. He suggested we camp under the cliffs at a rock overhang known as *yetiphu*. *Phu* in Sherpa-ka means cave.

We returned to Konar with the hay, and prepared to break camp in the morning. That night, in another part of Lama Sange Dorje's cave, the Down Jones closed 141 points lower, its third largest drop in history. The jitters were due to rumours that the US trade deficit for November would be the worst on record and that the government deficit would continue to grow in 1988. Meanwhile a presidential commission had found that one of the factors compounding the 19 October crash was that instant communications had created a global financial market.

We arrived at Macherma in thick fog late next afternoon. At one point Yangtse, wearing only tennis shoes and carrying the kitchen doko, took the wrong branch of the path and found herself on a slippery slope without safe issue. Pasang Tshering scampered up and took her hand to guide her back to safety, the two of them chattering and laughing all the way.

Four trekkers were in the kitchen of Macherma Lodge when we arrived: a Brit, an American Peace Corps volunteer, her Filipino boy-friend and an Austrian truck dispatcher. The lodge owner, Ang Zumbu Sherpa, was in Namche for market, but the lodge was tended by his sister, Karma Tangi, and a weathered yak herder by the name of Nursang. A sign on the wall read:

PLEASE USE THE TOILET
DON'T JUST SHIT EVERYWHERE OUTSIDE
AS IT MAKES IT UNPLEASANT FOR EVERYONE
THANK YOU
LODGE MANAGER

Nursang seemed a crafty, happy sort. He was curious to know what we were up to as it was unusual to see a single tourist travelling with a Sherpa family and their yaks. Moreover I had a pair of touring skis on my rucksack with 'Yeti' marked on them.

'The yeti skis?' he asked. Nursang knew about skis because one of his Khumjung neighbours, Lhapka Dorje, had fashioned a pair out of old barrel stays, put a wire binding on them and used to practise skiing on Macherma's south moraine. I had seen the tracks in April 1986 and was surprised that Lhapka Dorje had learned to make quite proficient parallel turns in unbroken powder. He apparently practised this sport, basically unknown to Sherpas, with the determination of someone training for the next Winter Olympics, which may have been his goal, until one day the previous winter he was swept away by a slab avalanche. He survived, but the wave of snow smashed him against a rock, breaking his thigh in two places. He was carried to Kunde Hospital where the Kiwi doctor patched him up. Three months in traction convinced him that the Winter Olympics could do without him.

Ten days before, Nursang told us, a yeti had come across the southern moraine from Luza, forded the stream and climbed towards the buttress of Macherma Peak.

Macherma's southern exposures are a favourite location for Himalayan snow cocks, which may have been what the yeti was after. They are pendulous birds that scratch at the sandy soil looking for food. They appear stupid and though naturally well camouflaged give themselves away by clucking when approached and then, rather than flying, they prefer to waddle uphill ahead of a pursuer in a test of endurance as to who will tire first. After photographing the snow cocks next morning, we left with Nursang to see the yeti tracks.

Nursang explained that the tracks had appeared one night when the moon was low. The yeti, he said, had descended the south moraine, after stopping under a rock overhang halfway down, close by the dark shoulder of Luza Peak. There was snow on the moraine's north side and with the binoculars I could see the tracks quite clearly. They mystified me. When later we walked alongside them, they appeared totally unlike the rounder Donag Tsho tracks or the stocking-footed 5,000-metre Knob ones. They represented another kind of 'yeti' footprint. Their form was elongated and narrow, more human in every respect than the others we had seen. In retrospect, they were similar to Tenzing Tsherpin's mummified foot. They measured about 24 centimetres (9.5 inches) long and 8 centimetres (3.2 inches) wide. The tracks were a single set, and the animal must have been moving fast, sort of loping along, for its flat-footed stride was more than 80 centimetres (over 2.5 feet) long. Although there were indentations for a left and right foot, ten days of sunshine had obliterated more interesting details such as toes, or heel. If Nursang had not told Gyalzen the animal was heading north, I'm not sure I would have been able to guess that from the tracks.

These were obviously not the tracks that Illinois Gary had seen, so I continued looking as I walked upstream towards our intended campsite at the head of the valley. About 100 metres (100 yards) from the smaller tracks I came across a triple set of markings in the snow that were so old they had been almost completely washed out. From what remained of them it appeared that they were similar in size and shape to those seen on the Knob more than a month before. They were roughly parallel to the smaller set, heading across the flat floor of the valley, which at that point was half a kilometre wide. For some of the distance at least they were three abreast, but because of their blurred state it was impossible to tell if they were heading north or south and in places the snow had completely melted so the overall impression was very confusing. They were, however, running in a more or less straight line between the network of caves on the north moraine I had already explored some weeks before, past a boulder the size of a small house in the centre of the valley, and the shoulder of Luza Peak with its overhang shelter under another huge boulder.

But why three sets of tracks? Nursang had returned to his yaks so we could not ask him. As I found no others, I supposed these were the tracks that Illinois Gary had seen and walked beside. At this stage it was impossible to tell which were yeti imprints and which might have been made by tourists, or indeed if all three had been made by booted humans.

Nursang had told us that every March or April, usually as

a result of a heavy snowfall, a solitary yeti descended from the high valley, where we were headed, but that its trail became lost in the lower valley. Two years ago, the valley was traversed by two yetis walking side by side. Above *yetiphu*, where we were going, was a summer pasture called Longsampa, difficult to reach and rarely visited. During the past five monsoons, according to Nursang, seven yaks were killed by yetis at Longsampa. I forgot to ask how he knew they were killed by yetis, but everything indicated we were entering promising territory.

Yetiphu was at the upper end of the valley, under the ramparts of Macherma Peak and near the base of the cliffs that blocked the passage into the high valley. With Marsan and Tsao we had good bait. Water was nearby, a small trickle between the rocks, though the trickle froze solid at night. Being at *yetiphu*, perched at an altitude of 4,750 metres (15,604 ft) within the ramparts of Macherma peak, was like being on a balcony with a view of the scree valley below, the hanging glaciers of Kyajo Ri opposite, and on our right the cliffs — which Gyalzen said were called 'yeti village' because they were pockmarked with not easily accessible caves. The hamlet and pastures of Macherma were out of view to the left.

We camped here for four days, exploring the base of the ramparts where a series of overhangs provided regular shelter for animals. We found cakes of yak dung and pellets of tahr excretion. Some of the dung cakes had been overturned, as if to dry, but there was no sign of fire or other human presence. The caves of 'yeti village' were too dangerous to approach because blocks of ice and rock were constantly breaking off from the lip above.

On Tuesday, 12 January, after a breakfast of potato and mushroom stew, Pasang Tshering left for home to help his mother prepare for Lohsar. During the night Marsan and Tsao had wandered down the valley and so Gyalzen and Yangtse had to retrieve them. I said I would meet them at Macherma Lodge, 2.5 kilometres down the valley from our campsite, and took my skis to cross scree valley.

The snow on the south side of the valley, under the walls of Luza Peak and the ridge that connects it to Kyajo Ri, was mixed and basically unpleasant. When I got to the place where both the big and small sets of 'yeti' footprints crossed the stream I climbed diagonally up the moraine, zig-zagging back and forth to reach the rock overhang where Nursang said the yeti had rested. In places the snow underneath me sounded hollow, which meant I was on old slab, and that made me nervous. At the rock, about the same size as Donag's Rock House, I took off my skis to explore the area around it. There was a north-facing overhang and behind the rock a narrow cave that burrowed into the mountain. The cave did not look like it had been used, but on the sandy floor of the overhang was what appeared to be the imprint of a body and two furrows that looked as if someone, stretching out to rest, had dug in their feet so as not to slide downhill. I looked for hair, but found only a few small feathers and some bird droppings.

The weather was deteriorating, with snow starting to fall, so I skied to the crest of the moraine to see where the tracks crossed towards Luza. Higher up on the shoulder of Luza Peak I could see with the binoculars a much larger cave with tracks in front of it, but as it was now snowing quite heavily I gave up any thought of reaching it.

At the top of the moraine I found the big tracks, again two abreast whereas in the valley they had at times been triple. It was impossible to tell where they were going; the south side of the moraine was bare of snow. It was clear where they had come from, however — in an almost direct line from 5,000-metre Knob, 3.75 kilometres to the north. They had crossed Macherma valley and climbed, via the overhang motel, to my very feet. And if they continued in the same southerly direction for another 3.5 kilometres they would reach the forests of Dolle. The animals had travelled between salient features of the landscape that offered protection or resting place, such as caves, overhangs, and house-sized boulders.

These tracks, though old, were for me additional evidence that the Donag yetis — Monticule Man and his wife — were travelling creatures who in the winter went south. These tracks were distinctive. They certainly had not been made by yak or tahr, the two most common animals in the area. Given the routing, particularly because they went from one feature to another, they had not been made by Sherpa or tourist. They avoided villages and habitations, indeed they stayed as far away from the main north-south trail as possible, while remaining parallel to it. I had uncovered a Yeti Marg or well-trodden travel lane, with its motels, rest stops and spas. If Donag had been the starting point for this yeti couple and Dolle was their objective, then Yeti Marg extended over a distance of about 20 kilometres (12 miles).

I skied down to the lodge, where Gyalzen, Yangtse, Nursand and a hunch-backed Sherpa were watching me, to find that Omai, one of Parajuli's trekking sirdars, had arrived from Kathmandu with my mail. We had tea as the snowstorm gained in intensity and then headed back along the 2.5 kilometre distance to our *yetiphu*, which we reached after 3 p.m. with 5 centimetres of fresh snow on the ground and more still falling.

The storm had settled in for the night and we could hear avalanches coming down around us. The mountains rumbled each time an avalanche broke off and each time I wondered whether a rush of compact snow was aimed at our little camp, although we were well protected by the overhang. At one point I heard Gyalzen reciting prayers in his tent. By 4 a.m., when it stopped snowing, there were more than 15 centimetres on the ground.

The sun returned at 7.20 a.m. and treated us to a hauntingly beautiful vista of the mountains and valley cloaked in white under a navy blue sky. Everything was now still, though after the fury of the mountains during the night, I understood why yetis like to descend into the valleys when it snows. But no yeti tracks were in the valley beneath us, only those of a fox and weasel.

The footing was so slippery on the steep mountainside that we could only move off the ledge with difficulty. But Gyalzen and I decided to look for the route to Longsampa through a nearby cut in the mountain. We wound our way above 'yeti village', along a shelf that progressively became narrower but apparently was still yakable. Then there was a stretch of shale debris and finally we could see where it widened onto Longsampa. There were no tracks on Longsampa and as clouds had started to roll up the valley again, and with the footing treacherous, I decided this was not a day to take risks and we returned to camp.

With Lohsar, the Sherpa New Year, six days away, we broke camp next morning and began our descent. At 9 a.m. it was already incredibly warm and this should have warned me. I crossed the scree valley for the last time and put on my skis in the day-old snow. After encountering the tracks of fox and weasel several times, I began climbing to the top of the south shoulder. Three-quarters of the way up I stopped for a rest. I was admiring the snow-swept mountains when I suddenly heard a hissing noise. I looked at my feet, but my skis weren't moving. The only reason they weren't moving, I quickly discovered, was because my skis were on a magic carpet of snow that was moving, faster and faster, down the shoulder. I was, in other words, standing on a snow slab that had detached itself from the slope. I debated in a fraction of a second whether I should attempt to ride it out. The slab was about 10 metres square. Then I saw we were about to plunge over a lip so I bailed out, turning on the slab before it broke apart and sliding off to the left. It had already carried me about 20 metres.

Gyalzen, who had been watching the action from the valley, let out a yelp. I replied similarly so that he would know there was no problem and that, after peeling off my seal skins, I would be getting the hell out of there. I had learned by than that Sherpas often communicate with each other across valleys or from one side of a mountain to another by yelping or imitating animal noises. This was a system that Gyalzen and I had adopted when tracking or walking together. If enveloped in fog or climbing a ridge by different routes we could tell exactly where the other was by telegraphing these noises, letting each other know whether the going was difficult, whether there was danger, or something to see. That was a measure of how well we had gotten to know each other and it produced a satisfying feeling.

After lunch of noodle soup at the lodge we left for Dolle, crossing to Luza by the normal route. Three Sherpanis were sitting on the doorstep of one of the dozen rock houses when I passed, skis on my rucksack. No, they replied to my question, they had not seen any yetis. They laughed. But on the high crest of the moraine I could see the vestiges of a double set of tracks descending to a large boulder about three-quarters of the way down the slope.

Once beyond Luza I understood where the yetis that I imagined had come off the 5,000-metre Knob were heading. Sloping downhill from Luza towards the Dudh Kosi river was a trail that led to a stunted forest inhabited by musk deer. The forest became thicker, offering more cover, as the river descended towards the hamlet of Dolle. Seeing this, the pattern for me was now clear. From Dolle, my supposed yeti couple would be able to cross the Dudh Kosi to the hamlet of Konar, spend some time in the inviting forest by the river before climbing past the hamlet to Konar Pass. From the pass it was only an hour's descent to the rock camp belonging to Gyalzen's neighbour on the east side of the ridge. From the neighbour's high rock camp they could move at their leisure to the more tempting forests around Deboche and Tengboche and then follow the yaks north again when they returned to the high pastures in spring.

We stayed that night in Dolle with Lhapka Dorje's two sisters, Lhapka Degi, about 22, tall, good looking, and newly married, and Frua Sange, about 19 and full of bounce. No sooner were we settled by the fire than hot chang was served and yeti stories began to whiz back and forth across the open hearth. No yeti had been seen around Dolle since Ang Dorje, a 28-year-old deaf-and-dumb native of Khumjung came across one drinking from the stream at Tshzecho, an hour's walk away. That had been in 1979. Ang Dorje hit the yeti with his stick. Now legend has it that anyone who harms a yeti will die within two years. Ang Dorje, until then in prime health, succumbed eighteen months later, after being 'two nights sick'. His parents, now very old, lived next door to Lhapka Dorje.

Lhapka Degi had a small baby tied to her back with thick green mucus oozing from his nose. 'New man in the house,' Gyalzen had noted when we arrived. Watching Lhapka Degi prepare our dinner of dal bhat was an experience. She took a khukri and started chopping the last nuggets of meat from the snout of a yak's head which looked like it had been hanging around for weeks. The baby was trying to grab the snout from her and I thought he was going to lose his fingers. Gyalzen and Omai were ladling chang from a plastic expedition drum and Yangtse was chirping in a corner trying to act the role of a little woman, fascinated by Lhapka Degi, who never stopped joking or laughing.

We returned to Pangpoche next afternoon, Friday, 15 January. Kanche had already left for Namche market to do her Lohsar shopping and the house was spotless. As was my custom when spending the night at Gyalzen's home, I listened to the BBC World Service. The news from Washington was jubilant. The November trade figures unexpectedly showed a 25 percent improvement, rather than the record increase that had been rumoured at the start of our sortie, and the dollar was strong. The currency dealers obviously had pulled off a tidy coup. During one week of work they had churned out paper profits that would easily have bought everyone in Namche Bazar a yak herd or two and still left them with a lifetime's supply of chang money. At $13.5 billion, the November deficit still remained considerable, but it was at its lowest level since April 1987.

Chapter 17

Lohsar

Our original plan had been to accompany Omai as far as Tengboche on his return to Lukla, but on Sunday morning, 17 January, a child from the upper village arrived unannounced in the long room to invite us to a pre-Lohsar pujah that afternoon. This was the usual practice in Sherpa villages. Although preparations for the pujah may have been in progress for days, invitations were only issued the same morning by a child sent around to the houses of those selected to attend.

To confuse matters, Sherpa and Tibetan Lohsar — their New Year festivities — are celebrated a month apart. Sherpa Lohsar, known as Sonam Lohsar, is held at the beginning of the twelfth Tibetan month. Sonam Lohsar is for farmers because at Gelu Lohsar, the more traditional New Year observance, held on the first day of the Tibetan calendar, they are already preparing to cultivate their fields. In Sherpa country, Gelu Lohsar is nowadays generally observed by religious institutions only.

During Sonam Lohsar every family worships their clan god. A great effort is made to clean house, but above all a year's dirt is removed from the hearth, and flour paste is used to decorate the walls and hearth with good luck symbols for the coming year. Presents are exchanged, especially new clothing for the children, followed by drinking and feasting which usually continues for several days.

That afternoon's pujah was hosted by Nang Dorje, one of

the village's richer citizens. He was a successful yak breeder and trader and his big house was located near the gompa. Gyalzen gave me a *khata* with 20 rupees folded into it to offer him. Gyalzen and Kanche brought gifts of rice and raksi. We arrived about 1 p.m. and the lamas were chanting prayers at the far end of the long room where plates of rice tsos and tormas were elaborately displayed on the family altar. As at Gyalzen's pujah, women and small children were seated on the right of the long room, the men in order of social consequence along the windowed side of the room.

In pole position by the fire was the host himself. Next to him was Gyalzen's father, and I, apparently now qualifying as an ancient, was placed next to him. Gyalzen was on my left. As the prayers progressed, accompanied by a strong rhythm section of rattle-drums and cymbals, we were served *dudh-chiya*, salt *chiya* and chang, followed by lunch of noodles and meat, raksi and more chang. The food was being cooked on the porch by Nang Dorje's son-in-law, Sunderay, who is one of only two men in the world to have climbed Everest five times, on each occasion without oxygen. (The other is Ang Rita, a yak herder from Thame). Among the other guests was Nang Dorje's sister, Yanchin, owner of Above The Cloud Lodge at Lobuche.

Gyalzen explained that each month at least one Pangpoche family organized a pujah. Each pujah cost around 4,000 rupees ($160). The people of other villages such as Khumjung or Kunde were not as pujah-minded, preferring to spend their spare cash on improving or repairing their homes. Consequently, Pangpoche had the reputation of being a fun-loving place where the villages spent all their money on getting the gods and themselves good and properly drunk.

In the late afternoon the plates of *tso* were distributed, with suntallahs, bananas, peanuts, popcorn, chocolate, sweets, biscuits, pastry, and Sherpa doughnuts stacked around their base. The doggie bags made their appearance and everything quickly disappeared. After the *tso* plates, dal bhat was served with more raksi and chang. One of the village didis, an older lady by the name of Tsosang, was completely in her cups and started ridiculing Gyalzen for selling his soul to a tourist who wanted to disturb the yeti. She went to each of the seven lamas and mocked our intentions, then did a tour of the room cackling about the lunacy of it all. Nobody said a word, particularly not Gyalzen, but as Tsosang was staggeringly drunk nobody took her seriously. Still, it was embarrassing and at 6 p.m. we left for home, our Chamonix shopping bag full of *tso*-plate leftovers.

Little Mingma Ramu had a runny stomach and was crying. Gyalzen sent Kunga Pemba to fetch the village lama from Nang Dorje's pujah to say prayers for her. After a few quick mantras, the lama suggested he should return the next evening to see how she was and if necessary do a stronger pujah for her then. Gyalzen agreed and the lama hurried back up the hill to Nang Dorje's house. Nevertheless Kanch remained uneasy. Mingma Ramu's illness was worrying her and made her short-tempered.

Mingma Ramu was now a year old and the medicines I had, aside from asprin and Immodium, a Swiss pharmaceutical remedy for upset stomachs, were either too strong or not suited to the child's complaint and therefore I was reluctant to give Kanche anything more than vitamins for her. Since my medicine chest had been unable to produce a cure for Gyalzen, she no longer had much faith in the pills and potions it contained and preferred traditional Sherpa medicine. For Kanche was convinced that Gyalzen's health had been returned by the pujah-bu and not the medicine Wade Hendricks had given him.

Next day Mingma Ramu was if anything worse, so the lama returned for prayers, torma-making and rice-throwing in an effort to induce the bad spirits to leave the house. If Mingma Ramu showed no improvement in the morning, Kanche said she would call in the 'Sherpa doctor.' The doctor, a *shaman*, was supposed to be able to see bad spirits and ghosts that the lama could not.

On the eve of Lohsar, 19 January, a New Year party was in progress at Ang Dorje's house next door. Already stuffed with food and drink, and not wanting to be the target of more ridicule, I was not eager to attend. But Gyalzen paid his neighbour a short visit. When he returned he had a heated discussion with Kanche. Naively I told them I was quite happy to stay at home. But this was not the problem. Since our return from Macherma I had noted an uncommon degree of tension in the air, with Kanche pushing Gyalzen to do something, though from their discussions I could not tell what it was. Gyalzen now came over and sat beside me. He wanted to express himself seriously in English but was unable to find adequate words. He started to cry. He was worried, he said, about Pasang Tshering's education and talked about wanting to send him to school in Kathmandu. But he said the financial burden was too heavy for him. He was too old to return to expedition work and had no means of earning a living other than in the trekking industry, which was so essentially seasonal. Gyalzen, unlike his brother-in-law Pasang, had really sold his soul to the tourist trade and could not easily reintegrate into the traditional Sherpa ecomony.

Having sensed that the issue of Pasang Tshering's education would be raised sooner or later, I had already decided to assume some responsibility for his future studies as I considered he had above-average learning ability. He seemed more than willing and it was a shame to let his talent and energy go to waste. But I asked why he had not been sent to the Hillary secondary school between Khumjung and Kunde and made to board during the week with a local family. Pasang Tshering was very difficult, Gyalzen said, and did not think that the Hillary school was good enough for him.

I told him that I had my own financial problems and did not know that I could undertake such a serious responsibility but suggested that when we returned together to Kathmandu we could look at schools there and see what it might cost. This satisfied Gyalzen. 'Thank you, Father, thank you,' he kept on repeating. Kanche, whom he always addressed as *ama*, or mother, remained silent even though she was the instigator of this confrontation. Gyalzen was a man driven by his wife.

I doubt Gyalzen would feel upset if he knew I considered his tears to have been largely crocodile ones, but for him to go through all that rigmarole meant the issue was important. I only hoped that my premeditated decision to help would not have the same consequences as the owners of Trekkers Inn had experienced with their teenaged son after sending him to Kathmandu for higher education.

The Sherpa doctor, a portly matron in a blue down jacket, dirty orange woollen hat, traditional angi to her ankles and a dirty striped apron, arrived late that evening, accompanied by a shorter female assistant. They installed themselves by the fire and questioned Kanche carefully about Mingma Ramu's problem, from time to time looking me up and down. Then they talked between themselves, did an inspection of the long room, asked what was in the blue expedition drums, and sampled some food. They discussed with Kanche and Gyalzen the ingredients they would need for the magic potions they intended to prepare.

Gyalzen set to work making a dough chicken, colouring its feet, wings and hackles red. He placed it on a copper plate with some rice, a spruce twig and yak butter, and a peacock feather was stuck into the chicken's neck. Various cups and bowls containing water, chang, raksi, and different food such as jam and coleslaw were placed around it. Finally Gyalzen was asked to recite from the holy books and throw rice.

While this was going on everyone was served hot raksi. The *shaman* was a good psychologist, drawing everyone into the preparations and having them talk about what they thought the problem might be. In the end, Kanche, who was still on edge, gave a long outpouring. Gyalzen was quiet and withdrawn during this venting of emotions and the children's eyes were often on me.

Finally a small bowl of embers was brought back to flame, sprinkled with incense that Pasang Tshering wafted through the house and the ritual incantation began over the baby. The children snickered. Then the dough chicken and other offerings were taken downstairs and placed by the door. It was now 11 p.m. and Kanche started mixing sweet dough for

Lohsar pastries. The deep-frying of the dough rings continued through the night. The *shaman* and her assistant were given covers and a place to sleep on the floor, next to the children. That night there was no moon.

Wednesday, 20 January, was Lohsar. Everyone was served chang porridge for breakfast and given a plate of Lohsar treats: suntallahs, peanuts, biscuits, cake and doughnuts. 'Today we eat many meals,' Kanche explained.

Little Mingma Ramu was said to be much better, but after a while it became apparent that she was not. A child came to invite us to Lhapka Tenzing's house, a few doors away, for yet another Lohsar party. When we arrived a bit before noon the party was already in progress. Among the guests were Sunderay, his diminutive wife, mother, sister and brother-in-law, and many of the same guests that had attended Nang Dorje's pujah. But it was much livelier. There was a lot of good-natured talk around the room, laughing and some singing — forlorn-sounding songs that Sunderay's sister began and everyone joined in. Four generations were present. The seating was traditional: women on the right, men on the left. Almost all the women were nursing mothers.

During a break outside for fresh air and a leak, Sunderay came over and, after some introductory small-talk, cautioned me about disturbing the yeti. Most people in the valley knew by then that I was looking for the creature. Even one trekker we met had greeted me by name. 'You're quite famous in these parts,' he had informed me.

'The yeti is like god,' Sunderay said. 'Mitey you cannot see. Chuti you can sometimes see. But if you see it, you die.'

I was not sure when he told me the mitey was invisible whether he meant that I could not see it because I was not Sherpa and therefore by definition a doubting outsider or whether he truly believed it was like a spirit. Many Sherpas, after all, had claimed to have seen the mitey and for them it did have a physical presence. But as Sunderay's English was about as good as my Sherpa-ka, I was unable to draw him out on the matter.

Sunderay was like a bear, with huge shoulders, a deep rib

cage and arms almost to his knees. He had thick black hair and small, light-brown eyes. With Ang Rita he had become a celebrity in Khumbu, one of the dearest of the native sons. We gazed at the top of Everest. The day was springlike, without a hint of wind to whip a plume of snow off its summit. 'The gods must be drinking chang,' he said, motioning towards Chomolungma's peak.

'Why?' I asked.

'No wind on the summit,' he answered. This was a rare event that only occurred a few days each year.

'What's it like up there?' I asked again.

From the south summit, he said, there was a half-hour descent before a tedious two-hour climb to the Hillary Step, a snow-covered rockband about 100 metres below the main summit. This was the most critical part. If there was no crust on the Step, the snow would be like sugar, making it impossible to get a footing. 'When snow is like that, you might as well go home,' he added. Once past the Step the route to the summit lay along a ridge so narrow that there was not room for two people to pass and a single false step meant a fatal plunge on the right 3,000 metres into Tibet or on the left 2,000 metres into the Western Cwm.

Every time he had reached the summit, Sunderay said he made an offering to the Mother Goddess and thanked all the gods for watching over him. 'As long as you do that, everything is much safer,' he explained, adding that one had to be careful not to offend the gods.

The party went on well past midnight with much dancing and singing. Little Mingma Ramu's condition worsened during the night. She had a high fever and was crying. A more powerful lama came to say prayers. The ritual tormas were again made and the sacred texts again read aloud, punctuated with frequent and hearty belching by the lama. This time the central figure in the forest of tormas was a tsampa doll dressed in a strip torn from Mingma Ramu's clothing. Kanche's temper was becoming more deeply strained by the baby's illness.

Friday, 22 January, was Kanche's Lohsar party. The children were sent around the village to invite the guests, but most people were either too tired, unwilling or otherwise unable to attend. This was a big disappointment for Kanche. The only guests to turn up were Sunderay, his mother, eldest sister, brother-in-law, wife, and Lhapka Tenzing with his wife. To begin with we were served several glasses of chang eggnog. Gyalzen sat on his bed at the far end of the room, contributing little to a warming of the atmosphere, but gradually Kanche's chang started to take hold.

Sunderay had special charm. He smoked and drank, but at thirty was convinced it could do him no harm. He had dropped out of school after his father's death to tend the family yaks and look after his mother and sister. Although married for ten years, he and his wife had no children. His mother wanted him to give up climbing, but he said that the money — about 200 rupees ($8) a day — was too good to turn down. He remained deeply religious and believed that if he prayed ardently enough, the gods would always protect him. We danced and sang until late in the evening and then everyone stumbled off to bed.

The *shaman* and her assistant returned after the last guests left to continue their treatment of Mingma Ramu. A new development in the household had to be weighed and perhaps accounted for in the renewed efforts to coax the demons to leave the infant alone. For Lohsar, Pemba Kunga had been given a black puppy, blind in one eye. He was called *Keytipe*, which literally meant 'small dog'. Because Keytipe belonged to Kunga, Tsheden had treated us to a succession of tantrums, though at least for the moment she refrained from tormenting it, afraid the dog might bite. But only too soon she would learn she could boss it around, too.

For two days Pasang Tshering had been hammering away at a 4-litre yellow soybean oil container. I didn't realize that I was the intended beneficiary of this industry, for he was transforming the tin into a brazier to heat the lhang. He proudly produced it the morning after Lohsar full of burning sticks and it rained ash upon everything, forcing me to flee to the long room where Tsheden was busy screaming and trying to attract attention.

Later in the morning, Pasang Tshering was sent into the lhang to sweep out the debris from the Lohsar party. He swept it under my bed. A mouse had moved into the chest with the religious robes. This did not bother Gyalzen or Kanche. The doglet had taken to crapping on the floor in front of the lhang entrance and peeing on the banquette by the fireplace. If somebody thought about it, doglet made it outside to the back porch. But nobody really cared whether doglet made it or not. Sometimes Tsheden made it outdoors, sometimes not. Again, nobody cared. Keytipe was fed leftovers on the floor and milk was poured for him into a dent in the floorboards. That morning Kunga received a good swift kick from his father for playing hop-scotch in the house. The cold weather had returned and I was chilled to the bone.

Bijaya Kattel had sent a note inviting us to spend a few days with him in Phortse. I told Gyalzen I would leave for Phortse in the morning.

'Good idea, Father,' Gyalzen agreed. 'I come one day after.'

Chapter 18

Musk Deer

The five and a half kilometre trail from Pangpoche to Phortse soars a good 300 metres above the right bank of the Imja. In places it is quite steep, sometimes perilously narrow, certainly no broader than the width of a yak. The Sherpa 'engineers' who maintain and repair the local trails, filling in washouts or building by-passes around landslides, have constructed crude stone staircases to help weary travellers or yak caravans over some of the more abrupt passages. Now and then a traveller might have to stop while a yak, ruminating in the middle of the pathway because it is the only level ground, slowly rises and makes way. Walking fast, the journey takes two hours — at a mean 4,000-metre altitude — though afterwards you feel it in your legs and in your chest.

I could imagine, then, the state of exhaustion that Bijaya Kattel had felt after walking for fourteen hours over similar terrain, stopping only once for food. That had been eleven years before, when Bijaya, beginning his career with the National Parks Department, was the newly appointed warden of Rara National Park in western Nepal. He was telling me the story of this trek now because he said it had changed his life and sparked his interest in the Himalayan musk deer.

We were sitting on the first-floor porch of Nima Tenzing's

house in Phortse, looking south towards the mountains of Tramserku, Numbur and Kongde Ri and the cleft in the Imja valley where Namche Bazar nestled, hidden from our sight. The porch was bathed in sun and it was pleasantly warm for a late January morning.

In November 1976, Bijaya had joined a twenty-man posse to search for four village headman suspected of organizing a poaching ring that had killed hundreds of musk deer. Once plentiful throughout central and eastern Asia, the musk deer has become one of the world's most endangered species. For centuries it has been hunted and killed because the male secretes a pungent substance that is probably the most expensive animal product in the world, more expensive even than the horn of the *Rhinocerus unicornis.*

The posse, led by the divisional forestry officer, included seven armed policemen. They had seen evidence of the poachers' destruction during the two-day trek from the district headquarters at Jumla, 300 kilometres west of Kathmandu. Coming off the 4,570-metre (15,000 ft) Pugma Lekh mountain on the second day of the trek, they had found a line of snares that ran for several kilometres through the oak forest. The posse's destination was a mountain village of about five hundred inhabitants. Illicit musk deer poaching had become the village's major source of income.

Bijaya said that they had hoped to take the poachers by surprise, but after walking since sunrise, finally reaching the village late at night, they found that their arrival was expected. The posse was roughly handled by the villagers. There was a scuffle, shots were fired, the policemen were disarmed and the forestry officer almost lynched. They were held overnight in a hut, given no food or water, and only managed to escape the following evening after word of their plight had reached Jumla.

Twenty-four villagers were eventually arrested and the Kalikot poaching ring, one of the largest ever uncovered in Nepal, was dismantled. Its leaders received five-year prison terms. Later referred to as the Kalikot incident, Bijaya said it had left an indelible mark on him and gave impetus to his career. He has since become one of the world's leading experts on the highly-prized but disappearing *Moschus chrysogaster*, the name given by science to the Himalayan genus of musk deer.

'The incident became so deeply rooted in my conservation philosophy that I decided to do everything I could to protect the animal from extinction,' he told me on the second day of my stay in Phortse.

With the two helpers, Beg Bahadur Baniya and his uncle Chet Bahadur Baniya, Bijaya had only recently moved into Nima Tenzing's house, near the school at the top of the village. Nima, a yak breeder of considerable skill, slept on a bench in the kitchen while Bijaya and the Baniyas shared what normally would have been the long room except that in Nima's house it was square. The room was crowded with carved wooden shelves that held several metal chests, Nima's copperware, a few books and layers of folded blankets. A large tea churn was by the kitchen door, separated from the square room by a wall of rough wooden planks, and the terrace where we were seated was at the eastern extremity of the house, connected to the square room by a dark and dusty corridor.

Gyalzen and I in our travels that winter had often observed the doe-eyed musk deer as it browsed at altitudes between 3,400 metres (11,200 ft) and 4,500 metres (14,800 ft). It seemed incongruous that so dainty an animal could be so cruelly hunted by poachers. Its pelt is unexceptional, almost mousey in colour. Furthermore, it is one of the smallest members of the deer family — an average adult weighs only 10 kilograms (22 pounds) and stands about the height of a greyhound — and seems to go out of its way not to be a nuisance. It is unobtrusive and trusting, as it allows people to approach within metres before bouncing off through the trees.

Bijaya's knowledge of the animal made him sound like a walking encyclopedia. He mentioned that other than their diminutive size, musk deer have three distinguishing features, all associated with the male: they carry no antlers; they have instead sabre-like canine fangs; and the male also has a pod-like abdominal gland that secretes the highly aromatic musk. And it is precisely for this moist, grainy substance that the animal is hunted.

For centuries musk has been prized as a scent stabilizer. Even though chemicals have largely replaced it in the perfume industry, and even though this is the era of laboratory-bred antibiotics, musk is more valued than ever as a no-substitute ingredient of more than two thousand Oriental medicines. The main importer is Japan. Japanese pharmacists choose to ignore that dealing in the substance is proscribed under the 1973 Washington Convention on International Trade in Endangered Species of Wild Fauna.

In spite of the proscription a single musk pod sells in the local black market for about $50 in current 1988 prices, almost equivalent to one-third of an average Nepali mountain family's annual income. This means that by poaching three pods a year, a farmer could double his family's living standard. But if he sold three pods a month, by local standards he became a wealthy man.

'For every buck killed for its pod, another five females are usually caught in the intensely cruel snares,' Bijaya explained. So when a whole village is in on the poaching it means that several thousand animals are slaughtered before the area is finally poached clean.

The snares are workmanlike devices, crude but efficient, and they cost nothing to make. A ringal-grass cord looped in a slip-knot is attached to the top of a green sapling that is bent double and tied to the ground. When an animal trips the cord the sapling springs erect. The snare's loop catches it by a leg, leaving the animal to a miserable fate, dangling above the ground.

In the autumn of 1986, Bijaya said that Lhapka Degi, sister of Khumjung skier Lhapka Dorje, observed three Tamangs acting strangely in the woods around Dolle. Tamangs are not particularly liked or trusted by Sherpas, though there are exceptions. These ones came from Dhading district, to the west of Kathmandu, and they were clearly trespassing on Sherpa territory. When Lhapka Degi found some well-camouflaged snares in the woods around Dolle, she reported it to her neighbour, the uncle of park warden Nima Wanchu. The neighbour informed his nephew and Nima Wanchu sent a ten-member national park protection unit to investigate. The unit, which included both Baniyas, tracked the poachers to a forest camp near Dolle where they found parts of thirty musk deer carcasses.

One of the poachers escaped his pursuers by fording the Dudh Kosi river towards Konar. At this point the Dudh Kosi rages through a steeply treacherous ravine. Few animals attempt to cross it, but this Tamang had succeeded and, knowing he was safe, coolly sat on a rock to watch the capture of his two associates.

They were cornered atop a large boulder by the Baniyas. One Tamang surrendered. The other drew his curved khukri and had to be disarmed by Chet Bahadur, who himself carried no weapon but is a towering giant of a man.

'There had been two previous arrests of poachers in Sagarmatha National Park,' Bijaya continued as we sipped our salt tea, waiting for Chet Bahadur to finish cooking our lunch of dal bhat over a wood fire in the kitchen. 'One involved a Tibetan merchant. He was caught at Jorsall with sixty-nine pods on him. The other was an isolated case near Pangpoche. The Dolle incident was the first time in the park's history that poachers were actually nabbed red-handed.'

The two Tamangs, it turned out, were brothers. They never disclosed the identity of the third poacher. They were taken under escort to Namche. But where the trail is most precipitous the poachers tried to escape. One fell to his death; the other was recaptured and brought to trial. He received a three-year prison sentence. He told the magistrate who sentenced him that he was paid 1,000 rupees ($40) for each pod by a man in Lukla. He said he knew the Lukla man by sight but not by name. That man, the local kingpin of the illicit trade, has not been apprehended and presumably remains in the game.

Poached Nepalese musk, according to Bijaya, continues to make its way, through Bangkok, to dealers in Hong Kong who sell it to international buyers for $50,000 per kilogram, or $1,420 an ounce — more than three times the price of gold. With such strong incentive to trade in musk, Bijaya said he believed the only way to stop the slaughter was to farm the animal commercially and perfect a system for extracting the musk without killing it.

Musk deer ranching is not a new idea. The Chinese, heavy musk consumers, have experimented with it since the 1950s using a smaller genus known as *Moschus berezovski.* In 1979, Bijaya visited three musk deer farms in Szechuan Province. While female deer used for breeding were kept in relatively spacious pens, he said that the males were confined to small cages under what he considered inhuman conditions.

Returning from China, Bijaya started searching for a sponsor and finally in 1985 he interested Bruce Banting, director of the US branch of the World Wildlife Fund, in his idea of ranching musk deer in Nepal. As very little was known about the habits and habitat of the Himalayan musk deer, the US council of the World Wildlife Fund decided that the animal's biology had to be better understood before thought could be given to underwriting a ranching project. The council agreed jointly to fund a biological research study with the King Mahendra Trust, provided KMT would administer the project.

'Kattel's project is extremely important,' Hemanta Mishra had told me in Kathmandu. 'We want the mountain people to think of musk as a renewable forest resource. Our objective is to teach them how to farm the deer in the wild. This is one way of demonstrating that wildlife conservation is a two-way street from which they, too, can benefit.'

Kattel, who was then thirty-five, married, and with two children, seemed to me typical of Nepal's rising middle class. He had saved and scraped to get the best education available in his field of animal biology, receiving a master's degree in wildlife biology from Colorado State University, and now he was gathering material for a doctoral thesis on musk deer. He was absolutely determined and dedicated, viewing with pride what had been accomplished throughout Nepal in terms of wildlife conservation. A measure of his determination was that, at one point in his youth, doctors predicted he would never walk again. He suffered from osteomyelitis and as a child had had fourteen bone operations. Now here he was climbing up and down the mountains around Phortse without sign of infirmity or fatigue.

Contrary to the Gurung police officer we had met at Trekkers Inn in Namche, Kattel had not come to Khumbu as a *rhomba* — the Sherpa word for lowlanders — bent on imposing his ways, but as a student of Sherpa culture. He had learned Sherpa-ka and was careful to respect local custom. Nepali government salaries are not large, but he was paying for four Phortse children to attend secondary school in Kunde and his rented quarters were at all times open to any Phortse villager who sought his advice or who otherwise took an interest in the musk deer project. Quiet-spoken, observant and friendly, he had been taken by the people of Phortse to their hearts though at the outset they had been uncommunicative, and mistrusting of him.

On Phortse's sandy shelf the villagers are limited to crops of potatoes, radishes and buckwheat. But, considering the altitude, the rhododendron and birch forests on either side of the village, dropping to the Dudh Kosi gorge 450 metres (1,400 ft) below are rich and hardy. Bijaya calls the smaller of the two the 'school forest', as it is near the village's mud-stuccoed stone school house. The trees in both forests are covered by a hanging green lichen which provides an eerie, almost fairy-like atmosphere.

At first the villagers opposed the musk deer scheme. They even refused to rent Kattel a house. They feared his research might harm the deer, thereby angering the local gods. Fortunately, Bijaya is intensely patient, part-psychologist, and equipped with the essential humour and insight to deal with reticent Sherpas. He fostered a friendship with Lama Jangpo across the way at Tengboche which improved his local rating. In the end, Bijaya convinced the citizens of Phortse that rather than harm the animals his project would help protect them.

In January 1987, employing some villagers as beaters and the two Baniyas as trackers, Bijaya captured six musk deer. He used coconut fibre nets woven for him by the elephant handlers at Royal Chitwan National Park in Nepal's southern Terai region. Each net cost a modest 4,000 rupees ($160). His success marked the first capture for scientific study of a Himalayan musk deer.

His technique for capturing the animals is simple. He stretches the 6-metre (20 ft) nets through the trees downhill from the animal, usually at the bottom of a gully. The unsuspecting creature is then driven down the slope into the nets.

Because of its large ears, curved back, longer hind legs and rudimentary tail, the musk deer looks something like an oversized hare. Its movement is a leap and bounce. It walks but does not trot. Although its shoulder height is around one metre (3 ft), it can easily jump twice its height and therefore in theory could bound over the nets, which were only two metres (6 ft) high. For some reason, the animal always bounds straight into them and gets bundled up as the nets collapse around it.

The netted animal is immediately sedated. Bijaya uses a cocktail of two drugs; three-quarters of a milligram of Rompun, a tranquillizer, and half a milligram of Vetalar, an anaesthetic, which he injects into the hindquarter. This usually knocks out an adult for between thirty and forty minutes. It is then given a name, measured from head to hoof, and a radio collar is attached around its neck before it is released.

The radio collars come from Telonics Inc., a specialized electronics firm in Mesa, Arizona. Each collar costs $300 and weighs 300 grams. The first animal to be collared was an adult male who lived in the school forest. He was named Nima Gawa. Bijaya estimated that Nima Gawa was about five and a half years old. He weighed at capture 9.5 kilograms, had a shoulder height of 54 centimetres, a chest girth of 58 centimetres, and a total length of 101 centimetres. His canine fangs measured 5 centimetres. Soon after releasing Nima Gawa, Bijaya netted two females: Ang Maya and Chhindy.

The project was interrupted when Bijaya fell ill. Returning to Phortse in December 1987, he found that Chhindy had been killed by jackals below the school forest and that Ang Maya may have slipped her collar, or she may have been killed. Her collar, tampered with but still functioning was found near Pangpoche.

After our *dal bhat* lunch, Gyalzen and I helped set up the fibre nets in the school forest, where Chet Bahadur had spotted a female that morning. As we walked along the path to the forest, three school-aged girls were busily preparing a make-believe snack by a table-sized rock. They called to Chet Bahadur: 'Achu Bazai, come and play house with us.' *Achu Bazai* is a mixture of Sherpa and Nepali which means Brother Grandfather.

'Achu Bazai, come and drink chiya with us,' the three of them called again. They had lit a tiny campfire under two rusty cans filled with snow and had set some broken plastic cups they had found in the woods on top of the rock. Although not more than ten years old, they were already wearing the long angi wrap-over dress used by all Sherpanis, with striped aprons around their waists, and they looked like busy little mothers scavenging about the woods for more discarded 'utensils' to play house with.

'Everybody loves Achu Bazai,' Bijaya said. 'He also has the eyes of a hawk and can spot a musk deer through the trees at 300 metres. Each musk deer is like another child to him.'

Everything went according to plan — or almost. The

female was driven downhill by a series of whistles. She hit the nets with a leap and bounce and was bundled up in them. But as Bijaya and the two beaters ran to hold her, she wriggled free and was off through the birches. The action was over in less than thirty seconds as a streak of greyish brown disappeared through the scrub, out of the forest.

Next morning we transferred to an area in the west forest that Bijaya knew was inhabited by two deer. Within half an hour Chet Bahadur had spotted a buck resting under a tree in a sunny patch. Success. Sano Kanchha, a two-year-old weighing 7.5 kilograms was caught, measured and collared. Before his release Bijaya scraped several grams of dark, purplish musk with the consistency of moist gingerbread from his preputial gland. Little did we suspect that less than a month later this perky little fellow who squeaked and cried for us would be dead. He was supposedly killed by dogs, but nobody was fully certain.

Like Nima Gawa, who we had observed in the school forest earlier in the day, Sano Kanchha had his own personality which he revealed to us over the few week before his death. He was cheeky and would forage close to the lower houses of the village, looking for shoots of grass and buckwheat. Nima Gawa was more sedate, almost tame. His collar irritated him and wore away the fur around his neck.

Each collar emits rhythmic beeps on a different frequency. Every day Chet Bahadur and his nephew telemetrically track five of the animals with a hand-held antenna. The rhythm of the signals indicates whether the animals are grazing or resting. By interpreting the signals and plotting the movements, Bijaya had by the time I visited him developed a mine of information about these engaging little creatures. Most surprisingly, he said, is the tiny size of their home range, averaging 4.5 hectares per animal.

'This is at least three times smaller than previously imagined,' he explained. 'They are sedentary, not migratory, and remain on their home range the year round.'

'That makes them easy prey for poachers,' I remarked.

'Of course,' he agreed. That was one reason why the musk deer population was rapidly decreasing.

Bijaya had also observed that the musk deer mates between November and March. The buck makes a low rasping noise as he chases his wife. She responds with a mee-ing sound, as if she is half-protesting, half-encouraging the assault. The chase never lasts long. The female quickly submits. So far Bijaya had only found evidence of single births.

Males fight to the death if one infringes on another's territory during the mating season. Their fangs are razor-sharp. In the forests around Phortse, Bijaya calculated the population density was forty-five animals per square kilometre, with a ratio of just over two females for every male.

'The animal is a selective feeder — a nibbler. It selects towards more nutritious food,' he continued in his clipped encyclopedic style. 'Its favourite meal is the green lichen that abounds in the Phortse forests. But it also likes rhododendron buds, radish leaves, potato flowers and buckwheat shoots. These it ruminates during daily rest periods starting around 9 a.m. and lasting until mid-afternoon. Its sense of hearing is highly developed and it sneezes as a warning to others when it senses danger.'

To reach the lichen they savour, musk deer are efficient tree climbers. They have remarkable balance and can walk along tree limbs to nibble at and retrieve the lichen. They are most active in the early mornings and late afternoons.

Kattel's next step will be to develop a management strategy for farming the animal in the wild. He believes that each male can produce as much as 20 grams of musk per year, for a current market value of almost $1,000. This was twenty-five times more than the Lukla blackmarketeer paid for a single pod. Production of 20 grams per animal per year was also twice as much as the Chinese harvested at their ranches in Szechuan. The animal's peak period of production is supposed to be between three and eight years of age, but he thought it could continue to be an efficient producer up to the age of fourteen. Its probable lifespan is between sixteen and twenty years.

Gyalzen and I quickly calculated that one male could produce about $12,000 worth of musk in its lifetime. Gyalzen noted that not even being a trekking guide paid as much. He was impressed.

Kattel foresees the day when Nepal will become the world's only exporter of farmed musk, as all Chinese musk production is used for that country's domestic pharmaceutical market. He also firmly believes that the development of musk farming in Nepal will end the international trade in poached musk, thereby saving the animal from extinction.

Chapter 19

The Woodcutters of Namche

Nima Tenzing and his brother Mingma Tenzing had been married to the same woman for more than twenty-five years. But the brothers had a falling out and now they lived separately, Nima Tenzing in the house he partially rented to Bijaya Kattel, and Mingma Tenzing with the wife and three children from their joint marriage in a slightly larger house less than 200 metres away. That weekend the youngest daughter, Kanche, was being married to a Phortse boy. To leave Nima Tenzing full use of his house during the wedding ceremonies — they were being held at the brother's house, but Nima was responsible for the overflow — Bijaya had decided to go to Saturday market in Namche with the two Baniyas, Gyalzen and myself.

The night before in Nima Tenzing's kitchen Bijaya and I had talked until midnight about an incident that had occurred the day before I had arrived in Phortse. Some villagers had discovered the remains of two Himalayan tahr in a seldom-visited lower pasture by the Imja Khola, near where the bridge carrying the trail to Tengboche crosses it. The dead tahr were part of a herd that Gyalzen and I had seen coming down from Pangpoche. The herders who found them reported that the hindquarters of both animals had been partly eaten. Because of this, Bijaya speculated that they had been killed by a snow leopard. I thought that this seemed unlikely, as the only snow leopard tracks seen by Gyalzen me had been at higher altitudes, around Donag and Tshola Pass.

'Then what else could it be?' he asked.

'How about a yeti?' I suggested, thinking of my Monticule Man and his mate. If after leaving Dolle the two of them had crossed Konar Pass it could be that they were somewhere in the vicinity. And while I quite willingly believed that the yeti scavenged off the kills of his hunting mate and companion of the wilds, the snow leopard, I had also come to the conclusion from listening to the stories of the Sherpas that the yeti was not entirely dependent on that animal for his share of wild game. No doubt this also was in some lesser measure the case of Ubeidiyan Man when he roamed the Jordan Valley 1.4 million years ago. That is to say, while Ubeidiyan Man scavenged off the kills of the sabre-fanged tiger, he had sufficient cunning to hunt smaller animals on his own without the tiger's help.

Even I had discovered that I could get to within ten to twenty metres of the trusting tahr, and I was sure that a hungry yeti had infinitely more patience, more stalking skill and more power than I. Compared to a yak, which is somewhat more robust, the tahr would be relatively easy prey for the yeti. If, as Sherpas claimed, yetis could break the necks of the not-too-fleet yak simply by grabbing its horns and twisting, they could do the same with a tahr.

The idea of a yeti being the culprit behind the killings of the two tahr threw Bijaya and he wanted to know what made me consider such a possibility.

I told him about the encounter I had at Donag Tsho with a nocturnal visitor and the tracks I followed south from the 5,000-metre Knob. He was intrigued by the fact that the tracks had been heading towards the deer-inhabited forests at Dolle. I also told him the story of Gyalzen's neighbour who owned a rock camp under Konar Pass, and that if what this fellow had said was correct, the yeti, or more likely a pair of yetis, maybe even a family of them, were roaming the Imja valley right now.

Bijaya was not a total sceptic. If the yeti attacked yaks there was no reason for it not to go after tahr. Gyalzen and I had counted four herds of tahr on the flanks between Pangpoche and Namche, so it would have been a rich hunting ground for the omnivorous yeti.

'But does the yeti exist, Robert?' he asked seriously.

'Something killed those seven yaks at Longsampa,' I replied.

'A bear?'

That the big yeti was in reality a Tibetan bear, an animal sometimes known to walk upright on its hind legs, is a theory often expressed by yeti sceptics. It is a convenient theory because it would help explain the strange tracks which people like myself have occasionally photographed and until there is conclusive evidence to the contrary it remains a possibility.

'All right,' I admitted. 'Suppose the large yeti is a bear, which I personally doubt. Then two questions remain. Have you ever heard of bears killing yaks? I'm not saying it doesn't happen, I simply don't know. And also, what about the smaller yeti – the mitey? Too many Sherpas have encountered the mitey to claim it doesn't exist.'

Bijaya had no definite answers to these questions, and neither did I. I was curious to know, however, if there was any nutritive value in yak dung. He didn't think so, but was puzzled as to why I asked. I told him about the overturned dung cakes under the rock overhangs near *yetiphu*, above Macherma, and wondered if they had been put there to dry by a yeti who was laying in a winter's supply of yak-dung biscuits. No way, he said. Excretion could be used as fuel or fertilizer, but definitely not as food.

This was not exactly true, as some animals, such as dogs, seek minerals and vitamins in the faeces of other animals. According to the findings of another Russian scientist, Prof. Jeanne Josefovna Kofman, who in the 1960s researched the existence of wildmen in the Caucasus, this is also the case among some primates. In the summer of 1966 members of her research team uncovered the presumed lair of a wildman which contained a cache of food, including four round pellets of horse dung. The locals claimed that it was well known that wildmen were fond of horse dung. Prof. Kofman concluded that this was because of the dung's salt content.

Since we were on the subject, I asked Bijaya whether yeti droppings could resemble yak dung. As I had not been able to find anything I could identify as yeti faeces I thought maybe I had been looking for the wrong end-product — that it really existed before my eyes, but in a form other than what I imagined — similar, say, to yak dung. To this the reply was obvious. The shape and consistency of an animal's excretion would depend on its diet. If a yeti ate the same food as a yak, its droppings might appear similar, but then that would mean that the yeti was a grazing vegetarian and not a killer of yaks, tahrs or musk deer. Fair enough.

I had by then seen Bijaya bolt a radio collar around the neck of Sano Kanchha and knew it would be nigh impossible for a musk deer to slip its collar. Who, then, had removed Ang Maya's collar? Certainly not a jackal. A poacher, maybe? Or perhaps a yeti?

Our inability to solve these mysteries was frustrating and in my sleeping bag that night, stretched out on a broad bench in the square room, I lay awake pondering them and listening to the chanting, clashing of cymbals and drum beating coming from the bridegroom's house at the far end of the village.

Next morning the chang ceremony, an integral part of the wedding rites, was getting under way as we left Nima Tenzing's house. After a night of feasting at the home of the groom's parents, the groom, accompanied by his family and friends, filed through the village, everyone dressed to the nines, carrying a large barrel of chang which was presented to the parents of the bride outside the bride's house. Before accepting the groom's chang, the bride's mother offered everyone a cup of her own chang, and only once a guest had drunk from the cup was he or she admitted across the threshold and allowed up the stairs to the long room where the marriage pujah was about to begin in earnest.

After watching the exchange of chang, Bijaya, the two Baniyas and myself descended to Phortse Tenga, where we crossed the Dudh Kosi and began the traverse to Sanasa, the

Tibetan marketplace. Gyalzen had returned to Pangpoche overnight to attend another pujah and said he would catch up with us along the trail. When we reached the chorten above Trasinga, we looked back and saw a wobbling figure climbing towards us, gaily singing and completely schnozzled. The pujah in Pangpoche had been a good one, lasting till three in the morning. Then, passing through Phortse, Gyalzen had allowed Nima Tenzing to pour a skinful of wedding chang into him.

At Sanasa we stopped at the teahouse owned by warden Nima Wanchu's wife, a round, amusing lady with one child who crawled between our feet and another on the way. Gyalzen almost fell asleep drinking another portion of chang. Nima Wanchu's wife, meanwhile, filled up a plastic container with some of her home brew and gave it to Chet Bahadur to carry to her husband at Mendalphu Hill.

The Baniyas did not drink alcoholic beverages. They had asked for fruit juice, served to them by Mrs Nima's young Sherpani helper from a plastic bottle. When the bottle was empty, the girl threw it beyond the wall on the far side of the trail. Everywhere in Khumbu this was the Sherpas' solution to waste disposal. At Gyalzen's house the solid waste disposal unit was behind the wall at the end of the potato patch. Invariably, in the absence of real toys, Tsheden and Tyrant or one of the other children would drag some of the discarded items, such as film capsules, old batteries or empty vitamin containers, back into the house to play with.

In all Khumbu, however, Mrs Nima's garbage tip was the worst I had seen. Bits of plastic, old bottles, empty cans, polythene sheeting, tin foil, toothpaste tubes, pill separators, the imported waste of an affluent consumer society, were scattered on the slope at the far side of the trail and extended downhill for a considerable distance.

Bijaya and I stayed a while longer with Mrs Nima, giving the two Baniyas and Gyalzen a head start. The next part of the trail was where the Tamang musk-deer poacher had fallen to his death so we told Chet Bahadur to take care of the stumbling Gyalzen. When we caught up with them a half-hour later, Gyalzen was nowhere to be seen. We searched for over an hour. The fall-off to the Dudh Kosi gorge was particularly steep. We walked back almost to Sanasa, checking the downside of the trail, but could find no sign of the Pangpoche pundit. Next, we started checking the upside of the trail and found him passed out on a grassy knoll, snoring blissfully. Achu Bazai had to virtually carry him down the remaining half-hour to Pasang Lodge on the saddle by Mendalphu Hill, where we paused for more chang and fried yak meat. Night had fallen and it was now very cold.

While enjoying our snack, Pasang, the lodge owner, told us that two days before he had sent the two Sherpani helpers he and his wife employed to gather wood in the forest on the far side of the Nangpo Tsangpo, the river that rises under Nangpa Glacier and flows past Thame into the Dudh Kosi. When they got to the forest, under Kongde Ri, the girls heard what they thought was the chattering of a yeti. They became so alarmed that they dropped their dokos and fled as fast as they could back over the 3 kilometres of trails to Mendalphu Hill.

As we had arranged to meet Kanche, who was coming from Pangpoche with Kunga and the yaks for Saturday market, at Namche Bazar Lodge, Gyalzen and I continued down into the village. When later that evening we told Pasang's story to Sonam Girmey, the owner of the lodge, he thought it not at all surprising. It was well known, he said, that yetis frequently visited the forests under Kongde Ri. In wintertime they descend Kyajo valley, coming from the fringe pastures around Longsampa, west of Macherma. They usually cross the Nangpo Tsangpo near Phurte and climbed to the ridges of Kongde Ri. Sonam Girmey also told us that sirdar Ang Rita from Thame had lost a yak to a yeti at Bhote, near Phurte, three years before. Bhote, he said, was another favourite yeti haunt.

But Sonam Girmey's strangest tale that evening involved another Thame herder. It seems that one September evening a few years back this man and his son were cutting hay in the pastures of Bhote. In the tall grass they came across the freshly killed carcass of a Himalayan tahr. Unable to resist a slab of free meat, they cut off an untouched shank, hid it under the hay at the bottom of the father's doko in case they met a park warden on the way home, and set off for Thame. It was not very long before they noticed that they were being followed by a yeti who presumably wanted his piece of tahr meat back. The father jettisoned the doko and ran with his son the rest of the way home. The most striking thing, said Sonam Girmey, is that some months later the boy took sick and died.

Next morning we joined Bijaya, Nima Wanchu, the chief of police and other officials on the top terrace of the market-place to watch the exchange of goods and flow of commerce. The weather was fine and a little later we sampled the momos at the Café du Marché before climbing back up the hill to Sonam Girmey's lodge.

Over a glass of chang in the lodge's front yard, Nima Wanchu mentioned that the American Embassy had requested assistance from the National Parks Department in locating a Peace Corps volunteer who had disappeared during the year-end holidays. The young woman, who spoke fluent Nepali, had set off alone on a trek to Sagarmatha National Park. Nima Wanche had asked his staff to check the entry records at the Jorsalle gate but could find no trace of any such person having entered the park.

We later heard that the young woman had been last seen in the village of Manidingma and probably was murdered there. Robbery was said to have been the motive. Returning through Tengboche some weeks later we met Josh, another Peace Corps volunteer who had been a friend of the missing woman. Josh said she was carrying little cash and robbery, therefore, seemed an unlikely motive. Her body was never recovered and nobody, as far as we ever heard, was charged with the crime.

The merchants and lodge owners of Namche are reputed to be among the wealthiest in Nepal. And yet they are too busy making money to repair or restore their religious monuments. 'Why should we, when a UNESCO official will come along one of these days and offer to restore them for us?' a Namche trader reportedly told Brot Coburn, the American project co-ordinator for a mini-hydro plant being built at Tengboche. The implication of this statement was that earning money had become more important to the Namche merchants than earning sonam. This was on the face of it quite a serious development. The cult of western materialism appeared to be triumphing over the power of Guru Rimpoche's vajra and the reality of Sanga Dorje's magic. Meanwhile, Namche's religious monuments stood neglected.

The old trail into Namche passed under a free-standing gate. It is common to find this type of structure, known as a *kani*, on the outskirts of villages in Solu-Khumbu. The interior walls and ceiling of a kani are usually painted with religious figures and mandalas. The old folk believe that these mystical paintings have the power to stop bad spirits, that sometimes attach themselves to travellers, from entering the village.

Namche's rotting kani has a corrugated roof and no mandalas. Beyond it is the seeing-eye chorten, its dome cracking, two rows of mani walls and five stone houses that contain water-powered prayer wheels, all broken and falling down, their paintings having faded away. The derelict waterwheels stand beside or astride the stream that issues from the three wooden spouts near the bottom of the village. Water that powers a prayer wheel is supposed to be clean and unpolluted. But the stream and the path beside it are littered with garbage. The path — once it must have been idyllic — is rarely used anymore because the Kleenex Trail now comes into Namche by the market-place. There is no kani up there on the Kleenex Trail, the new artery of trade and commerce, which may be interpreted that the Sherpa traders no longer care if arriving travellers carry bad spirits with them. Bad spirits have become part of good business.

We had a lunch of dal bhat prepared by Chet and Beg Bahadur in the national park compound on Mendalphu Hill. Even though the hill is 120 metres above the centre of Namche, a hydraulic rampump propels water up there from the Dudh Kosi so that those barracked on the hill — police, army and National Park staff — do not have to descend to the three spouts for their water. When constructing this water system, the government also installed a second rampump that distributed water in three points in the higher reaches of Namche. One fountain, for example, was near Sonam Girmey's lodge, and another was just under Pasang Lodge.

The government turned over the installation to the people of Namche, who then became responsible for its maintenance. The rampump broke down two years later, and since then nobody has repaired it. This was not due to lack of funds. The village panchayat receives its share of the market tax, apparently not inconsiderable. Sonam Girmey and Pasang are both wealthy businessmen. In addition to their lodges, they are considered among the best expedition sirdars in Khumbu and their services are in heavy demand. Sonam Girmey owns a large mixed herd of yaks and the hybrid zoms and zopiaks which he rents to expedition organizers and trekking groups to carry their baggage. Pasang has invested his profits in a second lodge at Lukla. Although a supply of water near at hand would make the operation of their Namche lodges much less arduous, as long as Sherpanis exist who are willing to hump 20-litre jerry-cans uphill they don't see the need for paying for the rampump's repair.

In a corner of the dormitory at Sonam Girmey's Namche Bazar Lodge is a brightly painted and well-polished prayer wheel. At six every morning while we were there the grandmother crept into the dormitory and started turning the wheel while mumbling mantras. Every revolution of the wheel clanged a bell which ruled out further sleep. I admired the prayer wheel's well-kept appearance, not at all like the dilapidated ones down by the kani-gate. And I could understand that if you were lucky enough to have such a lovingly-kept private wheel there was no need for the public ones downtown.

That Sunday, 31 January, the Super Bowl was being played in San Diego between the Washington Redskins and the Denver Broncos. Voice of America had been awing its

Asian listeners with tales of football folklore. I wondered what a farmer toiling in the wadis at the south of the Ganges thought when he learned that Super Bowl seats cost $200 apiece — more than an Indian or Nepali farmer earns in a year — or that one minute of TV advertising during the Super Bowl game sold for $1.3 million. We also heard that Americans spend $5 billion a year at fitness centres in order to maintain their body tone and work off the extra calories of an overly-rich diet. Then, too, we were told that an American businessman had paid $165,000 for a pair of low-heeled crimson slippers, size 5B, because Judy Garland had worn them in '*The Wizard of Oz*'.

I concluded that VOA broadcast these snippets from the American Dream in order to convince foreign listeners that the best base camp from which to begin the pursuit of happiness or the quest for a better standard of living was called Gringo-style Free Enterprise. If this was their message I hoped it wasn't getting through.

That morning Gyalzen and I had a bowl of porridge each for breakfast and we carried with us a package of glucose biscuits for lunch. Kanche, Kunga and the yaks had already returned to Pangpoche and we were going across the Nangpo Tsangpo to see if we could find any trace of the yeti in the forests under Kongde Ri. I took no rucksack, letting Gyalzen carry my two cameras and tripod in his. Although it was only 8 a.m., on the main trail west to Thame we soon became entangled in a lively procession of wood gatherers. At first there were a dozen of them, ranging in ages from twelve to seventeen. They were yelling and shouting, full of life and having a fine time in the fair winter weather. We were all heading for the same bridge over the Nangpo Tsangpo.

It was a 650-metre drop to the river, incredibly steep in places, but the wood gatherers, now more than twenty of them, all friends and enjoying each other's company, bounded down the trail like mountain goats with empty dokos on their backs. We crossed the river and began the climb through the woods to a plateau-like pasture known as Satarna. A stream ran through it and there was forest on all sides. Here, in the first sunlight, the wood cutters assembled, now close to three dozen, including three girls. They were a boisterous lot and of course any self-respecting yeti would have fled this onslaught, which it occurred to me, was the intention behind their din. While some sharpened their axes on whetstones near the stream, others played jokes on or otherwise teased the girls. One thing was certain: they didn't want their pictures taken. 'No camera, no, no, no!' they shouted when I tried to line a group of them up for a picture with Kongde Ri in the background.

We left them as they were about to break into little groups and set off to work separate patches in the forest. The patches were never too far distant one from the other, for there was safety in numbers. The girls, we noted, preferred to work alone, without the benefit of male company. We turned our attention to the ridge, and began our climb through a birch, rhododendron and spruce forest with scattered thickets of bamboo. As we climbed we could hear, far below us, the shouts of the wood cutters and the sound of their axes.

In the last layers of forest there was a thick snow cover where we found tahr, musk deer and human or other primate imprints. Leaving the forest around 4,000 metres, we entered a zone of thick brush-wood where we saw more suggestive imprints leading to a rock shelter, and above it a nesting place where, over an area of two square metres, the stunted brush had been flattened. There was no way of knowing what sort of animal had left the well-eroded tracks or made this nest. On the east side of the ridge, in a rockface, Gyalzen had spotted two tahr, but tahr do not nest. Snow leopards and, supposedly, yetis do, though I estimated that this type of stunted brush was too robust to have been flattened by a big cat, even a family of big cats. I looked for tufts of hair and droppings, but found none.

The higher we climbed, the steeper the ridge became and we could no longer hear the wood cutters. At 4,500 metres, the footing now unsafe, we stopped and set up the tripod. We had a dominating view of Namche, Khumjung, Phurte, the Kyajo valley, Nangpo valley, Ama Dablam, Everest and the Nuptse-Lhotse twins. Before we had time to take in everything that interested us, clouds began to rise about us and we were forced to begin our descent.

The forest was now quiet, as the wood cutters had started their journey home. But on the way down by a lateral path we could witness the massacre. Hundreds of freshly chopped tree stumps studded the hillside and felled trees lay on the leafy floor like cadavers. The technique was to circle-cut a tree and leave it. The trees would eventually show signs of ill health and, once ailing, could be felled more or less legally.

The slaughter of trees is prohibited by the National Park and Department of Forestry regulations. This restriction is supposedly enforced by park staff and the military police assigned to patrol the park. But obviously because the Kongde forest is so inaccessible it is not often visited. We had noticed many windfallen trees on the ground, which wood gatherers are entitled to harvest. But windfallen trees, if they have been on the ground long enough, rot and do not provide the same quality of firewood as circle-cut trees.

When we returned to Satarna the pasture was empty except for two pairs of impeyan pheasants. We were under the clouds now and at 3:30 p.m. it seemed oppressively grey. We stopped to rest and eat our glucose biscuits, seated on turf by the stream. As the clouds pressed down around us, I pondered what really had frightened the two Sherpani helpers from Pasang Lodge. Having now seen the wood gatherers at work and at play, I could imagine that the girls had been the innocent victims of a practical joke perpetrated by the boys. It was the sort of skullduggery that would delight them.

With the wood cutters gone, the still atmosphere combined with the closeness of the clouds, an abundance of game, the shelter of the forest and the nearby precipices made Satarna seem like an excellent setting for the yeti.

'Don't you feel the yeti's presence?' I asked Gyalzen jokingly, observing what his reaction would be.

He looked about us. 'Where?' he replied.

'Behind.' I pointed westward towards the edge of the

forest where there was the beginning of a track patrolled by two pheasants. Gyalzen became uneasy.

'Come, Father, it's time to go,' he prompted.

If it was not for the daily presence of the wood cutters, I would have happily stayed at Satarna for a week or more. Charles Stonor, who had arrived in Namche ahead of the 1954 *Daily Mail* expedition, did just that. With a guide he crossed the Nangpo Tsangpo and camped near here for a couple of nights. In those days the pasture of Satarna was deserted from October till April, when the yak herds returned. Under Kongde Ri's cliffs, amid the sage-green scrub, he came across some rather blunt tracks whose average measurements were 10 by 5 inches (25.4 by 12.8 centimetres). They were old, but nevertheless he and his guide were convinced that they had been made by a yeti.

Gyalzen was now in a hurry to leave, wanting to be back in Namche before nightfall.

Suggesting mischievously that we might linger longer, I reminded him that, for once, we were travelling light.

'Ah, but the yeti also travels light,' he replied.

We saw a musk deer on our way down to the river. On the left bank, the climb back to the Thame trail was tortuous, even without a rucksack. In one place, I had to use a couple of over-handed grips to pull myself onto the next shelf. But the wood cutters, with 30 kilograms of split wood in their dokos, did this climb almost every day. We encountered the first of them seated around a small fire under a rock overhang not far from the river crossing. They had propped up their dokos beside the trail and were smoking cheap Nepali cigarettes while playing cards. We passed two more card-playing, cigarette-puffing groups before reaching the first terraced pastures near Phurte where the main party was resting, tired and somewhat quieter than that morning, but still laughing and joking with the girls.

It was dark when we reached PK Lodge and stopped for a glass of raksi. Brot Coburn, who some Sherpas called the 'hydro wizard of Tengboche', had been by earlier in the day to install Namche's first microwave oven for Pasang Kami.

Like Sunderay, Pasang Kami told me he did not believe that the mitey could be seen. 'It's too god-like,' he said.

I was not sure what PK meant by this. Did he feel that a foreigner like myself was not a true believer and therefore unable to see the guardians of the god's hiding places? Or could it be that he was trying gently to tell me something quite different? His real message might have been: 'Don't disturb our yeti. We respect it, and really we want it left alone.'

In the morning we returned to Pangpoche to prepare for an assault on Lho La, the 6,000 metre pass into Tibet that lay to the north of Everest Base Camp.

Chapter 20

Lights In The Shadow Of Everest

Nạwang Tenzing Jangpo, a mild-mannered sage, is known by some as a 'renaissance lama' because of his modern ideas in a land that had changed little since the days of Genghis Khan.

A few years ago, Lama Jangpo decided that his monastery needed electricity. Otherwise, he reasoned, 'we will not be able to attract young Buddhists here to study. Without electricity, how can we compete with the lights of Kathmandu?'[1]

But hydro-electric projects are expensive and the abbot of Tengboche had no hidden coffers to dip into, no war chest stuffed with rupees. Monastic communities are not wealthy institutions. They are dependent for their well-being and support upon contributions from the faithful and other well-intentioned people.

The abbot, however, was determined that electric light should shine in the shadow of Mount Everest. As if by celestial design, one day in 1985 a team of Swiss engineers arrived at the monastery and conducted a hydro survey for the abbot. They found that Phunki Drangka, a mountain stream that collects the melt-off from the snows of Kangtaiga, had enough 'head' to power a 20-Kw turbine, small but sufficient to electrify Tengboche. The only hitch was that they said it would cost $130,000 to install.

During his thirty years as the intendant Rimpoche and

[1]My source for the High Lama's quote is Broughton Coburn.

manager of Tengboche's affairs, Lama Jangpo transformed the monastery into the Sherpa homeland's cultural and spiritual centre. He revitalized the monastery school, increasing the number of novices to the highest on record, built a Sherpa museum, and because of his interest in the arts, encouraged a new generation of naive painters to settle there, among them the thawas Phurba Sonam, Tukten Ishy and the younger Tinly Lhundub Sherpa. He is also well-known to the leaders of the many climbing expeditions that pass through the area, as he invariably invites them to dinner and blesses their endeavours. Occasionally he seeks out members of trekking groups camped on the monastery grounds to discuss the latest developments in their home countries, thus keeping abreast of what is happening in the world beyond the Himalayas.

Though he prides himself as being a humble scholar, he is a religious leader of great influence. When the King of Nepal, a Hindu, needs advice on matters affecting his Buddhist subjects he consults Lama Janpgo. If the abbot of Tengboche has a suggestion, the King never fails to listen.

Tengboche Monastery was founded seventy-five years before by Lama Jangpo's predecessor, Lama Chatang Chotar. He was acting with the blessing of the high lama of Rongbuk, on the Tibetan side of Everest, 40 kilometres away as the snow cock flies. In those days, monks made the three-day journey between the two monasteries by crossing Lho La, a 6,000-metre pass that at the time was relatively safe but today, because of the receding glaciers, is impassable to all but the best equipped climbing expeditions. A theory holds, however, that the pass is still used by migrating yetis.

Although not very old, Tengboche has nevertheless superseded its parent on the north side of Everest. This is because Rongbuk was destroyed by the Chinese in the 1960s, and though now being rebuilt, it will never regain its spiritual predominance as long as Tibet remains under Beijing's domination. The practice of Buddhism in Tibet had become subject to the whims of Chinese commissars who appoint lamas according to their political rather than religious merits, and who, in any event, have banned the search for reincarnate *thulkus*.

Lama Jangpo's adherence to established Buddhist teaching, combined with his progressive ideas, has made Tengboche the heir to Rongbuk's high monastic tradition. It is now one of only two monasteries to perform Mani Rimdu, which originated at Rongbuk, and Lama Jangpo's insistence that it be maintained with added perfection and brilliance has made it internationally renowned. Twice, Tengboche's dancing monks have been invited to perform in Japan.

As soon as you walk under Tengboche's free-standing kani you feel the natural aura of the site, a place where man must have worshipped long before the monastery was built. With the snowclad peaks of Kantaiga, Tramserku, Ama Dablam and Taboche towering around it, and the ramparts of Lhotse and Nuptse guarding the approaches to Everest at the northern end of the valley, it is like entering an awesome outdoor cathedral.

The kani is decorated with the mystical circle paintings and a row of benevolent deities that keep the evil spirits away. Next to it is the Supreme Enlightenment Chorten, a replica of the one that stood at Rongbuk. The ochre-painted gompa, centre-piece of the monastery, sits on a rise beyond the chorten, a jewel in the middle of this natural cathedral.

Tengboche is cloaked in the shadows of past centuries. Crossing its courtyard is like turning the clock back several hundred years. To be sure, some concessions have been made to the twentieth century. The crimson-robed monks have taken to wearing crimson-dyed ski jackets and the latest-style jogging shoes. They also have digital watches and battery-powered cassette players. The gong in the high gantry above the courtyard whose resonance commences and ends each day for the thawas, is in fact an oxygen cylinder. And the courtyard of the high lama's private quarters is littered with kitchen hardware from past expeditions.

These concessions aside, the basic lifestyle has changed little since the first Sherpa tribes arrived in Khumbu during the mid-sixteenth century. In the monastery's large communal kitchen, the rafters are soot-covered from the open-hearth fire that burns throughout the day. The firewood is harvested from the surrounding forests.

In addition to the sixty monks who live at the monastery, the site is visited each year by more than six thousand trekkers. Lama Jangpo allows them to camp in the yak pasture in front of the monastery. Otherwise there are the three lodges bordering on the common, one of which belongs to and is operated by the monks.

The campaign mounted by the high lama to obtain backing for the Swiss hydro-electric project showed just how wily he really can be. Below the monastery, perched atop its wooded ridge at an altitude of 3,870 metres (12,713 ft), are some of the most superb birch, rhododendron and fir forests remaining in Khumbu. Wildlife abounds on all sides to such an extent that Lama Jangpo had considered constructing observation blinds to rent out to the trekkers, thereby earning extra revenue for the monastery.

'Unfortunately,' Lama Jangpo wrote in a letter to the officials in Kathmandu, 'the forests in the area around and below the monastery are being cut at an increasing rate, as firewood has until now been the only source of cooking fuel available here.'

'We have always observed a policy of forest protection, and we have done everything within our power to conserve firewood. But now the area's forests are being further encroached upon by villagers from Kunde and Khumjung.'

A walk through the forests rising from the Phunki and Imja Khola rivers provides ample evidence of the diminishing tree cover. Stumps, both old and new, abound like gravestones and the axeman's tolling can be heard daily in spite of the ban on taking green wood.

This was the issue that Lama Jangpo seized upon in his search for financing. The consequences of erosion in the mountains, caused by deforestation, brings havoc downriver. The sediment swept into the Himalayan river networks from the denuded mountain slopes destroys farmland and dwellings in the plains. At least, reasoned Lama Jangpo, if electrical cooking facilities could be provided at the monastery, the need for firewood could be substantially reduced, the forest around Tengboche would be saved and there would be no more erosion.

'In fact,' he said, 'the only remaining need for firewood here would be for occasional use by individual monks in their quarters — a small percentage of the overall.'

This was Lama Jangpo's 'angle', because really his primary concern was to provide electric light for the monks to study and recite their prayers by, and also to attract new recruits to monastic life, thereby guaranteeing Tengboche's continued development.

Lama Jangpo's quest was referred to the King Mahendra Trust. One of the trust's basic aims is to promote the development of small cost-effective alternate energy sources to minimize fuelwood consumption.

King Mahendra Trust provided the seed capital to get the project going and raised the rest of the money from the American Himalayan Foundation, headed by San Francisco attorney Richard Blum, whose wife is a former mayor of San Francisco. Work began at the Phunki Drangka site in the autumn of 1987 after the traditional *serkim* commencement ceremony was performed by the monks standing stop Tengboche ridge, reciting prayers, tossing rice into the gorge below, and playing cacophonous music to exhort the gods to look favourably upon the project.

As there are no roads in Khumbu, everything that comes to Tengboche must be brought there on the backs of man or by yak, unless of course it is flown in by helicopter, of which there are few in Nepal. Soon after the serkim ceremony, six-metre long sections of penstock pipe, headrace conduit pipes, wire coils and building materials for the prefabricated powerhouse began snaking along the precipitous trails to the Phunki site, 300 metres (985 ft) below the monastery. The material was carried by sweating, grunting porters from the airstrip at Lukla. It was a dispatcher's nightmare. A major logistical triumph, for example, was when a dozen porters succeeded in lifting the 300-kilogram generator across the narrow, swaying footbridge suspended by cables over the Imja Khola.

On-site management was entrusted to Broughton Coburn, a former Peace Corps volunteer and Harvard graduate who knows Lama Jangpo better than any other westerner. 'He's a heavy, an empire builder,' Coburn once remarked of the lama. 'You can be discussing theology with him when suddenly he asks, "What makes an airplane fly?" It sort of throws you off balance.'

With his fluency in Sherpa-ka and knowledge of Buddhist teaching, Coburn, 37, from Seattle, Washington, became one of Lama Jangpo's closest confidants. The lama provided him with an apartment inside the monastery. The monks call him Brot. Aided by Ang Kanchi, a self-trained Sherpani paramedic whose brother is the monastery's chief accountant, Brot taught a handful of monks the rudiments of electrical engineering.

By mid January, Gyalzen and I, though we had often encountered the penstock pipe being portered along the trail from Lukla, had not yet visited the power-station site. The trail from Tengboche ridge descends through the forest to a construction camp, where we found Brot bringing Lama Jangpo up to date with the latest progress at the site. Lama Jangpo was accompanied by the high lama of Tragsindo Monastery in Solu who apparently was envious of the power project and wondered how he could get the government to provide one for his monastery.

With the two high lamas, the project engineer, and Lama Dorje, the construction foreman, we were taken on a tour of the site, ecologically sound and miraculously neat. Talking of miracles, Lama Dorje showed us the transversal beam in the small, prefabricated power house, which had become lost in transport, but one night while the workers slept was mysteriously slipped into place, all by itself and without anybody hearing a thing.

Lama Dorje, from the village of Ghat, near Lukla, is a man of charisma. This tough but ever-cheerful, ecclesiastic was wearing a badge featuring the Dalai Lama on the front of his woollen hat. Throughout the winter he supervised a team of fifteen masons, welders and carpenters constructing the power station, forebay tank and headrace conduit. Manual labour was provided on a voluntary basis by the monks who were also learning from Brot and Ang Kanchi how to wire the monastery and the tourist lodges.

The water to power the 20-kW turbine tumbles from the hanging glaciers of Kangtaiga, also known as Saddle Mountain because its summit resembles a Tibetan war saddle. The water is taken by a 210-metre conduit to a forebay tank designed to prevent flow-surges from disrupting the turbine. A 310-metre steel penstock pipe then funnels it downhill to the powerhouse, from where it is returned to the Phunki Drangka.

Walking from the forebay tank to the water intake bay, a Sherpa labourer carrying stones from the quarry on the far side of the stream grinned broadly at me and pointed to a cave in the base of the cliff that rises to Tengboche ridge. '*Yetiphu*,' he said.

We returned to the construction camp and were served lunch and milk tea. Lama Jangpo also mentioned *yetiphu* as a site where the animal was known in times past to visit. He said that one could always tell when the yeti had been there because the cave smelt badly for several days afterwards, particularly where the animal had urinated. As far as he knew, the animal had not been there for many years now.

Gyalzen and I visited the cave. In the rock corridor leading to it were a dozen tahr nibbling at the bushes. There were two caves in fact, one of them the biggest I had yet visited. No bad odours, though, and no sign of a yeti having been there, but heaps of tahr droppings.

We climbed back up the ridge and had chang at Gompa Lodge with Nang Wasser, the honourable lodge keeper. He told us there had been fresh yeti tracks seen at Macherma about a week before. But when we tied him down to the exact time frame it turned out to be the day we left Macherma and we were sure no fresh yeti tracks had appeared in Macherma's pastures that day.

Nang Wasser was originally from Khumjung, but his family had yak pastures at Macherma and he used to spend his summers there as a boy. He told us that eight years ago one of his family's yaks had been killed by a yeti up near 'yeti village'. We asked how he knew it was a yeti. Because, he said, he had seen its tracks around the dead animal. He indicated each imprint was the length of his forearm — a good 40 centimetres. He also said he saw 'yeti toilet', meaning the animal's stool. He claimed it was *white* and smelt very bad.

Construction of the power station and the wiring of the monastery and lodges was completed in April 1988. Lama Jangpo's five years of perseverance had seemingly paid off. In addition to his spiritual duties he presided over a three-man electricity board which set the rates for providing current to the lodge operators and the monks in their private quarters. The board decided on a flat charge of 150 rupees ($6) per month for each installed electric light. The lodges each have five low-wattage light fixtures, bringing their monthly charges to 750 rupees ($30). According to Brot, it costs 800 rupees a month to operate one kerosene lamp, so the advantages of electricity were quickly apparent to the money-conscious Sherpas.

Brot also installed an automatic switch which during non-peak hours diverted electricity to high-voltage cooking units. But as each unit draws between 5 Kw and 7 Kw of power, Tengboche is limited to no more than four of them. The generating capacity of the power station was itself limited by the amount of 'head' available on the Phunki Drangka, and the Swiss engineers had calculated that even with all the snows of Kangtaiga behind it, the Phunki could not sustain more than 20 kilowatts of power on a year-round basis. Because of this, Tengboche's lights were only turned on from six to ten each evening, during which time there was not enough power for the cooking units, let alone space heaters.

Of course, fuelwood for cooking and heating still consumes most of the trees from the forests. But lights shone in the shadow of Everest and this was applauded as a modest contribution to forest conservation in Khumbu.

The gods, however, were not pleased, or so it seemed, for instead of light in the shadow of Everest, the Tengboche power project brought sorrow and ashes. Nine months after the official inauguration, the monastery burned to the ground. I found out about it from Tashi Jangbu Sherpa, who runs Everest Trekking in Kathmandu. On 29 January 1989 I walked into the Bar National in Chamonix, a meeting place for climbers from around the world, and was surprised to find Tashi. He informed me he was in town for the annual meeting of the French Alpine Club's Himalayan Committee. But he looked downcast and glum. I asked if anything was wrong. He replied that he had just received terrible news. He had only moments before come from the Chamonix post office where he had telephoned to Kathmandu and was told that during the night a fire had destroyed Tengboche Monastery. There were no casualties, but everything inside the monastery — the many priceless relics and statuary, the Mani Rimdu masks and costumes, and the hundreds of volumes of religious texts — were lost for all time.

'It is a national tragedy,' he added, by then almost in tears.

I asked if he knew what had caused the fire, intuitively sensing the answer. 'A short circuit,' he replied. He didn't need to add that because of Tengboche's faulty wiring — provoking a cultural loss for the Sherpas comparable to the burning in 641 AD of the Alexandrian Library — progress in Khumbu, and with it forest conservation, had been set back a hundred years.

Chapter 21

Snowbird Dumps Its Garbage

The scale of pollution at Everest Base Camp, thirty-five years after the world's highest peak was first conquered by Sir Edmund Hillary and Sherpa Tenzing Norkey, is a major ecological scandal. Granted, the mess there is not a life-threatening matter — that is to say no one is likely to die from it — and therefore its dimensions are small when compared to the millions affected each year by deforestation in the Himalayas. But nevertheless it is a galling tale of disrespect by the climbing fraternity, of arrogant disregard for nature by men and women who evidently believe their personal conquests are more important then preserving the integrity of a unique natural site.

It is also a story of neglect by the Nepalese authorities, who require that expeditions clean up their base camps sites before leaving. This stipulation is written into the climbing permits issued to each expedition leader. The trouble is that the Nepalese government has no way of enforcing this requirement. There is a shortage of trained National Park staff, and liaison officers appointed by the government to accompany expeditions to their base camps rarely go beyond Namche Bazar, where the amenities include comfortable lodging, good food and electric light to play cards by in the evenings.

When finally the government is forced to do something, as inevitably it must, climbers will just as inevitably grumble about unnecessary restrictions of their rights and freedom, when in fact they will be the only ones to blame, having brought the restrictions upon themselves. A system of fines or a litter deposit, refundable if the base camp area is left unpolluted, has already been discussed. In December 1986, a Himalayan Environmental Protection seminar in Kathmandu attempted to draft an enforceable environmental strategy for climbing expeditions. But no decision was reached on how to cope with the problem.

The Everest Base Camp setting has to be seen to be appreciated. At 5,350 metres (17,552 ft), it is precariously perched among the scree hills and ice pinnacles on the floor of the vast Khumbu Glacier. Base camp is quite clearly an unwelcome intruder upon a world of harsh disorder and upheaval driven by the force of uncountable tons of compact ice moving inexorably down a raw and scarred valley where at one moment the air is so silent you can hear yourself blink, in the next a wind rises with such intensity that it sounds like an express train coming through a tunnel, and where avalanches thunder down slabs of mountain four times higher than the Eiffel Tower.

As it is near the upper limit of the glacier, base camp is surrounded by the highest natural amphitheatre in the world, with the hanging icefields of Nuptse on one side, the ramparts of Pumori on the other, and between them the Khumbu ice fall, descending like a crystal staircase form the Western Cwm. To the south are the silhouettes of Taboche and Tshola peaks. Everest, which according to the latest calculations is 8,863 metres (29,078 ft) high, is hidden from the base camp's view by the West Shoulder, rising to a height of 7,200 metres (23,622 ft).

Gyalzen and I had gone to Everest Base Camp to reconnoitre and if possible climb, Lho La, the pass that faces Khumbu ice fall. In former times, Lho La was the most direct route between Tengboche monastery and its parent Rongbuk, and the monks of Rongbuk had told early British expeditions that a yeti family lived near the snout of the main Rongbuk Glacier. Several expeditions had, in the past, encountered yeti tracks around their base camp tents. In the case of one Polish expedition, its members followed a dual set of imprints to the rockfall at the base of Lho La. Gyalzen said this was logical as it was well known that the yeti crossed back and forth over Lho La.

We climbed halfway up the pass but abandoned our attempt after almost being bowled off its face by falling rocks. We found no tracks and concluded that Lho La is now impassable to all but the best equipped climbing expeditions. This is because the Lho La glacier has receded so drastically during the past half-century that what was once a gentle slope is now a near-vertical wall with pendulous ice overhangs and broken seracs waiting to cascade down 700-metre (2,300-ft) chutes.

Accompanied by Pasang Tshering, Mingma Yangtse and two Sherpani porters from Pangpoche, one of them Sunderay's sister-in-law, we had walked from Lobuche on the previous day, arriving at the base camp too late in the afternoon to do any exploring. The assembly point for hundreds of epic climbs was abandoned for the winter, as it was early February, and we had the site to ourselves and two stray dogs that had followed us from Lobuche. There had been barely time to pitch our tents and prepare a dinner of dal bhat before darkness fell. But when I crawled out in the morning I had a nasty surprise.

I exited from the tent on all fours and the first thing to greet me was a frozen strip of cloth. I had put my hand on it by mistake. I picked it up and upon closer examination discovered it was a used sanitary napkin. Next, I found fistfuls of used toilet paper. Around my tent I also counted many scores of batteries, Kodak film wrappers and other brand-name garbage. Then, too, there were prescription drugs still in their dispensary containers and a broad array of human faeces. Although others before me had put tents on the level platform of scree on which mine was now pitched, the area surrounding it was quite literally a dump: filthy, disgusting, and, once the sun began to melt the ice into little streams of faeces-contaminated water, potentially a health hazard.

This, though, was only the beginning. A little further away, tin cans and bottles, plastic containers, tubes, canisters, cardboard boxes, wooden crates and a single oxygen cylinder were strewn about as if the site had been a battlefield. I almost expected land mines to explode as I surveyed this farrago of waste. In fact a little later I picked up a discarded flare device, examined it and tossed it away, thinking it a used battery. But no! Upon impact with the ice it started to smoke and then exploded, scaring the daylights out of me. We were grateful, however, for the wooden crates as they gave us a supply of firewood.

Scavenging has never been a big thing with me, but I was fascinated by the fact that so much of the trash seemed to be of American origin. So that no doubt remained about this, Pasang Tshering brought into our kitchen that first morning a miniature American flag and an unopened package of Beer Nuts he had found nearby. The makers of Beer Nuts proudly proclaimed on the package that their product was 'America's favorite — the crunchy, nutty sweet 'n' salty one-and-only natural taste.' This claim seemed mildly incongruous amid the crevasses and scree of base camp, but when sampled with tsampa porridge the Beer Nuts did provide an interesting supplement to our otherwise drab diet.

Pasang Tshering also found an American-made thermometer, which was broken, and a red plastic baseball bat, which the manufacturer had named the 'Big Bopper'. While I was wondering whether it had been passed off to the Sherpa who portered it to base camp as an essential piece of climbing gear, Pasang Tshering returned with a two-toned brown disk.

'What's this?' he asked.

I examined it. On one side it carried the circular inscription: 'Mayor Byrne's Summertime Chicago.'

'A frisbee,' I said.

'A what?' Like the Big Bopper, Pasang Tshering had never seen one of these before. Because it was cracked, it was not in flying condition, so I was unable to demonstrate it for him. 'Special American climbing gear,' I said. 'But broken. No good.'

'Oh,' he murmured, and off he scampered in search of more mysteries.

Yangtse was intrigued by the advertising put out by Wilderness Experience of Chatsworth, California. She had found a 32-page catalogue of Wilderness Experience's outerwear garments that had once been attached to an 'Ultimate parka, red, small – our warranty', and some promotional hype for Quallofil, 'the modern alternative to down'.

Yangtse was also into pharmaceuticals – because of the small containers they came in, which were useful for playing house, and she had found half a deck of cards.

Now the fiction is that climbing expeditions, being ecologically minded, dispose of their garbage by dropping it into crevasses, of which the Khumbu Glacier has a fair number, or by burning it. There was, indeed, a narrow but deep crevasse outside my tent, another two on either side of the kitchen, and mounds of semi-burnt garbage in the scree hills beyond it.

From the trash surrounding our two tents and kitchen I found it was possible to recreate the physiognomy of the expedition to which it belonged. This was one first and foremost an American endeavour. It included both men and women, and, judging by the manufacturers' warranties left scattered about, no effort had been spared to equip them. It had been an affluent expedition, both in terms of material and garbage. While it seemed that no cost had been spared to outfit it, every cost had been avoided in cleaning up after it. But still I did not know *which* American expedition was the culprit.

It was only later, rummaging through a discarded porter's doko, that I came across a stack of unused postcards representing the 'Birth of Everest' by artist Phurba Sonam, one of the Tengboche monks. The postcards had been overprinted with the green logo of the 1987 Snowbird Everest Expedition. Then it came back to me. Snowbird was a ski resort and beauty spot in the Rocky Mountains of Utah, owned by Texan oil millionaire Dick Bass.

The 1987 Snowbird Everest Expedition, backed by the

Dallas-based Bass with some apparent help from the Adolph Coors Brewery in Golden, Colorado, had attempted to scale the monarch of Himalayan peaks by the South Col route during the 1987 post-monsoon season. Three years before, Bass, then fifty-four, had become the oldest man to climb Everest and the only man in the world to scale the highest peaks on all seven continents. Bass himself was not a member of the 1987 expedition, but he had chartered a helicopter to fly to Lukla with the intention of trekking to base camp to be with his climbers for the beginning of their assault.

According to records filed with the Ministry of Tourism, which issues expedition permits to Nepal, the Snowbird leader was experienced West Coast mountaineer Peter Athens. An assistant to Dick Bass, Karen Fellerhoff, was in charge of base camp administration.

Hit by the record snowfall of 19 October 1987, the Snowbird expedition, like most other Himalayan expeditions that autumn, was unsuccessful. Planned as a tribute to the Utah resort, by the mess it left behind it contributed to its shame and the shame of its sixteen American members. Had they dumped the same garbage in the streets of Snowbird they almost certainly would have been arrested. But clearly they had no inhibitions about doing it in the remoteness of Everest Base Camp.

Not only did the Snowbirds leave close to a ton of unburied, unburned garbage at base camp, they also had the dubious distinction of introducing American-style graffiti to Khumbu. During my travels that winter I had seen 'Snowbird' daubed in paint on walls, bridges and mani stones. At the time I had wondered what it meant. I had also seen Sherpas wearing baseball caps marked 'Coors Light' and I had wondered what that meant.

The Snowbird's prosperity to litter was all the more surprising because it breached the climbers' code of ethics embodied in the Kathmandu Declaration of 1982. This declaration was adopted by a general assembly of the International Union of Alpine Associations and signed by member organizations in twenty-six countries, including the American Alpine Club.

Article One of the declaration acknowledged that 'there is an urgent need for effective protection of the mountain environment and landscape'. Article Three stresses that 'actions designed to reduce the negative impact of man's activities on the mountains' should by all means be encouraged.

There was no doubt that the Snowbird expedition ate well. Moreover, they had imported American foodstuffs when local equivalents were readily available at a fraction of the cost. And they drank well, too. Among the trash were dozens of Coors Light beer tins. The shining 12-ounce 'All Aluminiun Recycle' cans lay in the sun like so many dead soldiers with no hope of being recycled. But the supply of Coors obviously had not been sufficient, as empty bottles of Nepali-brewed Star Beer littered the scene as well.

Karma and his wife Yanchin, owners of the Above The Cloud Lodge at Lobuche, told us that every few days a yak train arrived from Namche Bazar loaded with more Star Beer for the climbers. The Snowbird team seemed to enjoy tippling, and usually they drank the best; Chivas Regal Scotch Whisky, Johnnie Walker Extra Special Black Label, Dewar's White Label and Veuve Clicquot Ponsardin 1982 Carte d'Or champagne. In addition, there were uncountable bottles of Nepali Khukri rum. Now this was only what appeared on the surface, as there was still much snow about.

And should these intrepid mountaineers have wanted to nibble while they drank, they had a choice of snacks: smoked oysters served on stone-ground wheat crackers or tins of broken shrimp. Pringle's Potato Chips had been brought all the way from Cincinnati, Ohio, and the Chef-to-Go smoked salmon was prepared by the Washington Trade Corporation of Seattle.

By the look of things, the Snowbirds had brought enough food to fill a supermarket. Their menus were sumptuous. Breakfast started with Tang. General Foods, its maker, advertised on the gallon-sized tins that Tang had been 'selected by NASA for use on all manned space flights since 1965'. Dozens of giant-sized tins attested to the fact that the Snowbird team had enjoyed the orange-flavoured drink as much as the astronauts. But I hoped that no discarded Tang tins were suspended in space. I felt mildly envious when I discovered they also had a choice of pre-sweetened cereals, followed by tinned sliced bacon, and pancakes that they could smother in Golden Griddle Syrup, or muffins 'with wild Maine blueberries'. I hadn't eaten anything like that for months and was beginning to despair at the monotony of Sherpa food. I found some Maxwell House instant coffee as well as filter coffee from the Thanksgiving Coffee Company of Fort Bragg, California. Wrappers advertising Low-Fat Milkman 'with the kiss of cream' had been scattered by the wind more than 200 metres away. Empty cases of Coffee Mate, Carnation Evaporated Milk and Carnation Hot Cocoa Mix were also in evidence.

For lunch, tuna fish seemed to be preferred: at least four different kinds. Dinner was more substantial. To start with, they had a choice of packaged soups. A main course might have been Country-Style Beef Stew with long grain and wild rice (boil in bag, ready in seven minutes), but there were a dozen other freeze-dried dishes that ranged from Beef Stroganoff to Rainbow Trout Almondine. In addition, they stocked two kinds of tinned fowl: Chicken Ready (one whole chicken, without giblets, packed in broth), and Nugget's Diced Boned Chicken. The Kellogg's Company was represented by their Crouttettes Stuffing Mix.

The Snowbirds had brought their eating culture with them in no uncertain terms. It didn't occur to them that Khumbu, like Idaho, was potato country and that Sherpas actually know how to mash or even deep-fry potatoes. Snowbirds' mashed potatoes were by Betty Crocker or a Safeway Stores variety called Town House. They also imported cartons of General Foods' minute rice when Nepal is a rice exporter. And if between meals the Snowbirds became hungry there was a wide range of appetite soothers: the inventory included Snickers, M&M King Sized Peanut Chocolate Cookies,

KitKat Wafers, Hershey Bars, Nutter Butter Peanut Butter Cookies, Lorna Doone Shortbread and Carefree Sugarless Bubble Gum.

Empty food and beverage containers, wrappers and packing cartons — a few of which, like the bag of Beer Nuts, were still full — were not the only garbage left behind. Among the boulders and scree were discarded mail, beauty products, pharmaceuticals galore, gas canisters, toothpaste tubes, cigarette packs, clothing and eating utensils. The American Snowbird did not travel light.

Mary Kay Brewster, one of the climbers who had the best chance of getting to the top, left behind mail from a boyfriend, Stephen Kanipe, of Columbus, Ohio. There was an accounting note which said 'Peggie Duscher owes a total of $2,880 – this at cost. The Rockwells owe an extra $1,200.'

Someone was reading *Cry of the Kalahari*, for its pages were strewn across the base camp. Whoever it was had obviously learned nothing from the authors' search for pristine wilderness in Africa. Mary Kay Brewster, on her second Everest expedition, was applying to medical school. In a discarded draft of her application letter she described her Himalayan adventures as 'once-in-a-lifetime opportunities'. They certainly proved to be opportunities to foul the Nepali wilderness.

Other artifacts included unopened packages of disposal hypodermic needles and syringes, and for the woman unabashed by leaving a trail of dirty nappies, an unused Sani-Fem Freshette. This plastic device, patent pending, enabled her to stand rather than sit in public restrooms. Besides making for sounder hygiene, standing makes visiting public facilities more esthetic . . .' the instructions for use stated. But there are no public restrooms at base camp. Or, looking at the problem from another angle, base camp is today nothing more than one large restroom — the highest public facility of the sort in the world.

Every indication spoke of a disorderly retreat off the mountain by the Snowbirds under treacherous weather conditions. But I was puzzled. I remembered that an Austrian expedition had been on the mountain at the same time. The Austrians, too, had been beaten by the adverse snow conditions. I had met one of them in Pangpoche, on his way back to Namche Bazar. During our five days there, I searched the entire base camp area but could find no trace of the Austrian stay at the highest restroom in the world.

Chapter 22

Above The Clouds

From Everest Base Camp we descended to Gorak Shep, a forlorn place if ever there was one, littered with broken glass, used toilet paper, tin cans from around the world and other refuse. We camped in the ablation valley north of the frozen lake, beside a two-door rock house that had been built many years before by Dawa Namge, Gyalzen's father. Behind the house was a boulder which carried the fading inscription *In Memory Of Mick Burke, British Everest Expedition 1975*, and on the last green patch before Khumbu Glacier swallowed the ablation valley, about half a kilometre to the north of us, was the field of memorials for those who had lost their lives on the mighty Chomolungma.

Shepti in Sherpa-ka means dead and Gorak Shep derives its name from a dead gorak found there when Sherpas first drove their yaks to pasture under Pumori. The dead gorak was certainly an ill omen, for the place bodes no good. Even the Sherpas claim the stagnant water in the shallow lake can kill you, as it almost did me, and the place offers no protection from the raw Tibetan winds that sweep down from Lho La.

Our interest in stoping there was the Kalapattar above our camp which merges into Pumori's south ridge, from where we hoped to view the Changri glacier network to the west, and the additional fact that it had been the site of the first Everest base camp. It was near here that Dr Edward Wyse-Dunant, leader of the 1952 Swiss Everest expedition, found yeti tracks and, more towards Lobuche, came across what looked like the sleeping quarters of several yetis. The tracks had come from Changri La, a high pass that crosses westward into the upper Dudh Kosi valley, more or less opposite Ngozumba Tsho. The Swiss also found what they presumed were yeti tracks on Gorak Shep's small frozen lake and Wyse-Dunant concluded that the animal was 'a wanderer, avoiding zones inhabited by men'. On a second Swiss Everest expedition later that same year one of the porters was attacked by a yeti at Gorak Shep, and his screams brought other Sherpas running to chase the hairy two-legged creature away.

We climbed to the base of the Pumori ridge and gazed down upon the rows of moraines on either side of us. It was clear from the description left by the Swiss that the glaciers had retreated by at least three to five kilometres since the area was first explored almost forty years before. This was confirmed when we descended to Kalapattar's south summit. As we arrived there a young Sherpa guide reached it from below, leading a party of five panting Australians. By the way he talked about 'my grandfather', I realized that he was Tashi Gompu, Tensing Norkey's grandson.

'Are you the man looking for the yeti?' he asked. Karma and Yanchin at Above The Cloud Lodge had told him that he might encounter the 'yeti man' at Gorak Shep.

Tashi, who was twenty-three, lived in Darjeeling where he had learned to climb with the Indian Mountaineering Institute. Tensing Norkey was thirty-three when he reached the top of Everest. Asked what it had been like on top of the world when he returned to base camp, the Tiger of the Snows replied: 'I immediately prayed to Chomolungma. I offered the goddess some chocolate and biscuits I had with me, burying them in the summit snow. Then I thanked her and the other gods for their great kindness and also I asked for their forgiveness.'

Tashi said that in 1953, according to his grandfather, the snout of the glacier, its southern limit before the ice became covered by debris, was literally at our feet, Now it was 4 kilometres to the north. 'It's shrinking fast,' he noted. The greenhouse effect had taken hold not only of America's Middle West but was melting ice on the roof of the world. As the glacier level sank, Gyalzen explained, the Khumbu ice fall, which he knew well, became steeper and more dangerous. If the greenhouse effect continued for another forty years, there might not be any Khumbu ice fall left at all.

I knew that Tashi's grandfather, Tensing Norkey, had believed in the yeti, but I wondered if Tashi did. 'Oh yes,' he answered, adding that he and his climbing companions had often heard the long, piercing whistle of the yeti around their summer climbing base in the Kangchenjunga massif, but that personally he had never encountered one. His maternal grandfather, who was the caretaker at the camp during the off season, had once come across a yeti at a distance of 15 metres. 'It picked up a rock and threatened to throw it at grandfather.' Tashi said.

I asked Tashi how he could tell the difference between the whistle of a yeti and that of a Himalayan tahr. 'Oh, easy. A yeti's whistle is not as shrill or as short as the tahr's whistle.' he replied.

Tashi was serious about the yeti. He said Sherpas believed so deeply in the animal's existence that no matter how exposed they became to outside culture. 'they will continue to believe in the yeti for another thousand years'.

I told him that in a thousand years the yeti had every chance of being extinct as there would be no more glaciers for it to roam, only stony roads.

We left Tashi and next day changed camp, moving west toward the base of Changri Nup glacier. The area was a vast desert of moraines, one after another, the size of mountains. I thought I would be able to ski up to the Changri La pass but this proved to be a bad joke. You can only climb the Eiffel Tower so many times in one day and then you call it quits. The moraines here at the junction of four glaciers were as tall, and they were arranged in never-ending rows, so after spending two days in them, and reaching the much diminished glacier cap, I decided enough was enough. I skied across the upland to the south and descended to Lobuche where a litre of Kanche's best raksi was waiting at Above The Cloud Lodge.

Lobuche, in spite of the inscription painted on a rock that trekkers should burn or bury their waste, is another garbage dump of epic proportions. There are two lodges there and only one neglected outhouse for the six thousand defecating trekkers who make it to Lobuche each year. The place may have been beautiful when Shipton and the Swiss, followed by Hunt and Hillary, arrived there in the early 1950s, but it is today an eyesore.

One might have thought that even the yeti would have been frightened of picking up some scabrous disease at Lobuche, but Karma and Yanchin believed a couple of them still roamed the area. At the beginning of the winter some tourists had reported seeing fresh tracks and when the wind blows fiercely from Tibet and Mother Goddess Chomolungma turns down the thermostat, the Sherpas still fear that yetis descend from the high passes in search of shelter.

When we arrived back at Pangpoche the following afternoon, Kanche had butter tea waiting for us in the long room. Chomolungma had turned down the thermostat, for it was very cold. While seated around the hearth, two policemen mounted the stairs unannounced and walked into the long room. One was armed with a semi-automatic. They asked a few questions in Nepali, looked at me and then left. Kanche said they were looking for illegal wood cutters.

Pasang Tshering and Yangtse distributed the loot from base camp. Kunga received a sawn-off deck of cards, an empty Pringle's potato chip tube and the toy replica of a Ferrari, Tsheden a grey plastic mouse, some syringes and two empty spools of surgical tape. Pasang took the Big Bopper out onto the back porch and cut off both ends to make a red plastic trumpet out of it. He had never seen a baseball game in his life. Kunga, who had never been in Kathmandu, had not seen a motor car before. But Tsheden knew about mice. Mingma Ramu was still ill and whimpering. Kanche had given up on the shaman, but had the lama in nightly at 30 rupees a time to say prayers.

In the four months that I had been living with the family I had watched the children evolve. Kunga was now a teenager who walked around with hands in pocket, the family's number two wood gatherer but number one poser. Yangtse, the water carrier, was also number one wood gatherer. Pasang Tshering had grown about two centimeters and was a little man. Instead of calling his father Papoo he called him Galzei, the diminutive of Gyalzen.

The three elder children shared in caring for Mingma Ramu. Kunga was especially lavish in his attention, sharing food by injecting partly chewed bits into her mouth. Pasang Tshering, when his mother was too busy, sprinkled water on the floor and would do the sweeping. His method was to achieve a general redistribution of the dirt, most of it going into one or two designated corners. With mother and Yangtse out looking for Marsan, who had wandered up to Bibre, the dish-washing chore fell to Kunga who received some unwanted help from Tsheden. At one point the Tyrant started hitting him with a knife. One whack from her obviously hurt and he threatened retribution. She immediately started to cry. Gyalzen looked sharply around, but said nothing. She had failed to bring punishment upon Kunga. As soon as Gyalzen was tending to the fire again, she resumed hitting Kunga with the knife. The remarkable thing was that Kunga never whacked her back. Finally she got bored and started tormenting Keytipe, the small dog.

One morning a few days later, Kanche was doing the sweeping, collecting the dirt and trash into a corner by the stairs from where it would be eventually shovelled out to who knew where. Tsheden came into the room and shuffled through the dirt in front of her mother's broom, then stopped to pick up two or three pices of trash she wanted to preserve — lengths of string or wool, a piece of plastic. Later, Pasang Tshering was seated in front of the fireplace drinking tsampa chang with his parents. Tsheden walked up behind him and for no reason thumped him as hard as she could. Not only did this surprise him, but it caused him to spill the chang. He told her to stop with remarkable restraint. She burst into tears and told her parents that Pasang Tshering was being mean to her. Pasang Tshering said nothing. The parents didn't react. Tsheden changed her tune to one of instant and adoring love.

On one of our last trips to Tengboche, we noticed a lot of musk deer fur on the ground as we climbed the ridge, by the only fresh-water spring for the three lodges and monastery. At Gompa Lodge a soldier was standing guard over the torn carcass of a young deer killed during the night, supposedly by stray dogs. Its chest had been ripped open, the rib cage smashed and the hindquarters were partly missing. Nang Wasser also told us that Sano Kanchhe had been killed by dogs in Phortse.

'Dogs or yeti?' I asked.

'No yetis here anymore,' Nang Wasser replied. 'They've gone away.'

If the yetis have gone, then logic would have it that the gods have gone too, chased away by the trekkers and the diminishing forest cover. For many, the tourist industry was bringing progress and development to Khumbu. The Sherpas, I was told, no longer suffer from goitres because a broader selection of foodstuffs was now available in Namche to meet the demands and tastes of the tourists. But, manifestly, progress had also brought pollution and a degrading environment. Because of the fragile eco-system, had Khumbu reached the saturation point of what it could safely absorb in terms of progress and tourism? Was progress chasing away the gods?

Interestingly, in the Himalayan kingdom of Bhutan, where untouched primeval forests still exist, the government recently began cutting back on the amount of development the country was thought to need and placed even tighter controls on tourism, estimating that the saturation point had, for the present at least, been reached.

At the Protection of the Himalayan Environment conference in Kathmandu in December 1986 there had been a timid discussion of restricting the number of trekkers allowed each year into the Annapurna Sanctuary and Sagarmatha

National Park. The trekking industry was dead set against such a move.

One fine morning towards the end of February, Pasang Tshering announce that he was not going to Kunde school, but had decided he would attend the same school as his two cousins in Bodnath. He claimed he spoke better English than they did, and this was qualification enough to be admitted. That evening he recited his prayers and read from the tantras in a louder voice than usual. We would leave the next day on the first leg of our final descent to Lukla.

On Wednesday, 24 February, Pasang Tshering was the first up. He lit the fire and made tea. By 8 a.m. he was ready to leave, striding about the long room and giving sharp orders to everyone. He was feeling his oats, indeed superiority over other members of the family. Gyalzen confided to me that Pasang Tshering's behaviour was becoming a problem.

The old lady Tsosang was the first to arrive with a jug of chang. She draped a white scarf around my neck and wished me *tashi delai* (good luck) and poured everyone a glass of chang. She said that Pasang Tshering's care and education was being entrusted to me and she lectured me about the seriousness of this undertaking. This from the lady who two months before had ridiculed our efforts to find the yeti. Ang Dorje and his wife arrived next with a jug of raksi, followed by a brother-in-law, Gyalzen's mother Passangma, Sunderay's wife and, finally, almost half the village. Except Uncle Pasang. Neither he nor his wife Nima Yanchin came to say goodbye.

That night we stayed in Tengboche and drank chang with the monks and Brot, the hydro wizard, for the last time. I was reminded of Lama Jangpo's theory that the alleged yeti hand at Pangpoche gompa was actually the severed hand of a robber. I asked Gyalzen if he believed this. Certainly not, he replied. Everybody knew it was the hand of a *zimbu.* I had never heard of a *zimbu* before, so I asked him what it was. Gyalzen described for me a human vampire which he said once existed in Tibet and fed upon the blood of children.

Whenever one was discovered, the Tibetans would execute it. Pieces of the zimbu's anatomy were then distributed to various monasteries and gompas to frighten away others of their kind.

But why, I asked Gyalzen, did everyone refer to Pangpoche's skeletal hand as a yeti hand? Ah, he said, because the caretaker of the gompa spoke little else but Sherpa-ka and when asked by visitors why the hand was kept in the gompa his usual response was the two words of English he knew that satisfied everyone: 'Yeti hand.'

That was another example of Sherpa logic. Moreover, had I asked Gyalzen if he was superstitious he would have ferociously denied it.

We made it to Namche late on the following afternoon. Kanche was already waiting there and she was carrying the rolled up Tibetan carpet upon which I had slept during the past four and a half months and she said tradition would have it that I continue to sleep on it wherever I went. She also had brought two Tibetan necklaces and a ring as her present to my daughters.

Needless to say, in Kathmandu we got Pasang Tshering into the Baba Boarding High School at Chabahil, near Bodnath, first kitting him out with school uniform and gear for a full eleven-month academic year. The head master, Mr T.S. Pradhan, estimated that Pasang Tshering was four years behind other pupils his age but was confident the young Sherpa would quickly catch up.[1]

I also went to see Hemanta Mishra and when I mentioned the Snowbird garbage at base camp he was saddened. He wanted me to provide him with pictures of the garbage so that he could show them to Dick Bass who was due back in Nepal for the inaugural flicking-the-lights-on ceremonies at Tengboche Monastery. He intended pressing Bass to remedy the situation.

---

[1]By December 1988, at the end of his first school year, Master Pasang Tshering was at the top of his class and Mr Pradhan reported that he had earned a 'double promotion.'

Some months after returning to Europe I received a postcard from Brot Coburn. It said: 'I thought you might be interested to know that Snowbird is considering donating $15,000 to a system for *perpetual* collection of garbage and good sanitation along the Everest trek route. Who else is doing this sort of thing?'

That was good news, but upon further reflection it struck me that this was merely a grandstand manoeuvre by Dick Bass. His gesture amounted to nothing more than a one-time publicity shot as it would only pay for a maximum of ten hours of helicopter flying time, hardly enough to lift Snowbird's garbage back to Lukla. So where was the perpetuality?

And what about the yeti?

Personally I believe the yeti is still out there, roaming the uplands with its mate in search of an undisturbed territory that together they might call their patch, with adequate food, shelter and the prospect of a secure future.

I have never talked to a Sherpa who doubted for a moment the yeti's existence. The yeti is so much a part of their lives that, ethereal or physical, it does exist. Their sightings of one or other yeti types are too numerous to discount, so its presence is more than ethereal: it must be real. Brot Coburn told me that on one of his trips into Tibet he took a tape recorder with him and conducted a survey in the villages he visited. He is fluent in Tibetan and soon realized that 'Have you ever seen a yeti?' was the wrong question. He began asking instead, 'When did you last see a yeti?' He said the detail people could remember about their most recent yeti encounter was astounding.

The Sherpas have developed over the centuries physiological characteristics that permit them to survive in an inhospitable climate at high altitude. They are short and stocky mountain people with a large chest cavity for more lung power and they have a capillary system that is much less sensitive to extreme cold than our own. It is, therefore, not impossible that a relict hominid, clad only in a thick mantle of natural body hair, could have evolved regressively to live an animal's existence in the high forests and pastures of the Himalayas. Nor is it impossible that a large gibbon or langur could have similarly adapted to a high mountain existence.

While our two-man expedition was not the success we had hoped for, it was by no means a failure. My travels with Gyalzen and Uncle Pasang established the presence of a large plantigrade creature around Donag and Macherma. The midnight tsonkers, the Rock House scamper, the footprints left by the Pond creature, the Monticule wanderers and the Knob couple, the falling rocks in the night on Donag's North Slope, Gerard's nocturnal visitors, the Kongde Ri and Pangka nests, the two half-devoured tahr below Phortse, and a musk deer's radio-collar found 4 kilometers from the missing animal's home range were, in my mind, no series of coincidences. Even the snow leopard has its part to play in the yeti mystery.

The 11-kilometre stretch from Donag to Macherma, where we observed most of our yeti activity, was the same area where the 1954 *Daily Mail* expedition had found a 'maze' of footprints and several bagfuls of droppings. The 1954 footprints were similar to those left by the Knob couple and the Monticule wanderers, whom I believe were the same pair of yetis, and their 'territory' was only 25 kilometers from where Eric Shipton had photographed his yeti footprints. Shipton's photos, and those taken by the 1954 *Daily Mail* team, show imprints remarkably similar to the tracks left at Donag by Monticule Man. There were also the footprints we had been unable to examine on Kongma La and Khumbu Glacier. They suggest, with those heading down Gokyo valley from Donag to Dolle, a seasonal migratory pattern over a set route in an established territory.

When George Schaller and Peter Matthiessen travelled into Outer Dolpo, a bulge along Nepal's northern border that erupts onto the Tibetan plateau, in search of the snow leopard and found none, they didn't claim the animal was extinct. The proof that it was extant came fifteen years later when American zoologist Rodney Jackson returned to Dolpo and became the first to radio-collar a snow leopard.

Jackson later visited Khumbu and couldn't find a trace of the snow leopard there, but it didn't mean the animal was not present in Khumbu, as Gyalzen and Uncle Pasang already knew and I soon discovered.

But if, as I maintain, the yeti resides in Khumbu, why have so few westerners seen it? After all, we have the accounts of yeti sightings by William Hugh Knight, N.A. Tombazi, Don Williams, Reinhold Messner, Gary Scott, and a handful of others. Pascal Marlinge's gangly man aside, only Scott's encounter with a pair of hairy yeti-like creatures was inside Khumbu.

There are three valid explanations for this. The yeti is by preference a nocturnal wanderer, except perhaps during the monsoon. But in the monsoon there are no climbers or trekkers around. The Khumbu may only encompass 500 square miles, but the immensity of that countryside can hardly be imagined. If all the mountains were flattened and the moraines levelled, Khumbu would probably be as large as Texas. And if its sparsely-inhabited immensity is a second factor for the lack of sightings, the roughness of terrain is the third. The difficulty of covering ground in yeti country is perhaps the animal's greatest asset in its quest for privacy. It can reach remote caves and hidden valleys where no Sherpa, let alone Caucasian, can follow without a major physical effort and often with considerable danger.

Even the glacial cut on Rawldurche that Uncle Pasang and I explored, possibly unvisited by humans for twenty or even fifty years, took us one half-day to reach, and a whole afternoon was required to descend its 1,500-metre length. There are thousands of such places, many even less accessible. No, I am certain there is something very real hiding out there and that one day it will be photographed, but by then because of the rape of the environment it may be too late to save it from extinction.

Early in 1989 I received a letter from Kanche. It was hand-carried to me by a friend who had passed through Pangpoche. It was a bit vaguely addressed. On the envelope someone had simply written for her: 'Father, Switzerland, Europe.'

In it she wished me 'hearty greetings' and mentioned that 'we think of you from our hearts always.' She also reported that Gyalzen was on his way to Gokyo with a trekking group, but there was no reference to the yeti. Still, it started me thinking, with the result that I decided in 1990 to return to Khumbu and have another look for Monticule Man and his mate.

Almost half a century after Eric Shipton's 1951 encounter with the Menlungtse footsteps there is no clearer, more succinct argument in favour of the yeti's existence than the now famous Appendix B of MOUNT EVEREST 1938, written almost light-heartedly by his friend and climbing partner H.W. Tilman. In it Tilman suggests that if fingerprints were sufficient evidence to hang a man, then surely footprints are adequate proof of an animal's existence. The footprints certainly are there, because I and others have photographed them.

While the Yeti '88 'expedition' did not find an Abominable Snowman, it increased the general body of knowledge about the animal and in the process destroyed a number of myths. According to the Sherpas, Tibetans and other Himalayan people, there are two and probably three different species of yeti-like creatures. The *migo* (or wildman) aside, about whom very little is known, the other two — *mitey* and *chuti* — are territorial, but of no fixed abode. They are not, as most people assume, solitary creatures. Frequently they move in pairs, sometimes with their young. They follow a pattern of transhumance, descending from the high pastures with the first snows to roam the forests of Khumbu's lower valleys during the harsh winter months, slowly beginning their ascent again in the springtime. Their range lies between 3,400 metres and 5,000 metres: the same as yaks and Himalayan tahr.

Contrary to the often-stated fallacy that a large mammalian creature would not have enough sustenance to survive, there is sufficient food in those high-altitude regions to sustain a modest yeti population. They are, moreover, nesters, and move freely from one site to another along established travel lanes, harvesting what the land has to offer along the way. They are omnivorous, part-time scavengers. As such, they share similar behavioural traits with *Homo erectus* or perhaps even early Neanderthal Man.

They are definitely endangered and must be protected. Their forest habitat is under siege and their high-pasture ranges are increasingly invaded by trekkers and climbing expeditions. Because of these intrusions, their sightings around such 'tourist' meccas as Tengboche have totally ceased.

Perhaps most interesting from an anthropological point of view is that Sherpa and Tibetan legends about the yeti are similar to legends about the sasquatch that abound in the cultures of west-coast Indians in Canada and the United States. North American Indians and Himalayans believe, for example, that contact with the big-footed animal brings bad luck, even death. Some Red Indian tribes, as some Sherpa clans, claim that the yeti/sasquatch will attack and even devour humans. All accord it the respect due to a semi-sacred presence and none dare harm it.

I know of no other person who has established, through personal observation and research, that the yeti moves along travel lanes. In the Khumbu, one lane descends the west side of Gokyo valley to Dolle, and from there probably crosses Konar Pass to the Imja valley. Another parallels the east side of the Imja valley from Khumbu Glacier to Mingbo and Deboche. But others certainly exist. In any event, I am probably the only person to have followed the tracks of a pair of yetis over a distance of 20 kilometres. This convinces me that yetis continue to roam those scant forests and high, rocky pastures. But for how much longer? That is the question.

Appendix A

Sponsors and suppliers who contributed to the Yeti '88 Research Project:

Nikon AG, Zurich, Switzerland, cameras and spotting scope.

Carl Zeiss, Oberkochen, West Germany, binoculars and Orion nightscope.

Phoenix Mountaineering Ltd., Amble, Morpeth, Northumberland, England, makers of the world's finest tents.

Dynastar Skis, Sallanches, France, manufacturers of Yeti touring skis.

Essex Chemie AG, Lucerne, Switzerland, distributors of Solarex Ultra Sun Screen.

Nic Impex, Annecy, France, distributors of Sun Pocket heaters.

Revue Thommens, AG, Waldenburg, Switzerland, makers of the Thommens TX 9000 altimeter and precision compass.

Téléphone S.A., Geneva, Switzerland, communications equipment.

Nestlé S.A., Vevey, Switzerland, suppliers of light-weight foods, Fitness chocolate and condensed milk.

Ciba-Geigy, Wander and Hoffman La Roche contributed pharmaceutical products, protein supplements and vitamin compounds.